THE COUNTRY ALMANAC OF

HOUSEKEEPING TECHNIQUES

THAT SAVE YOU MONEY

First published in the USA in 2012 by
Fair Winds Press, a member of
Quayside Publishing Group
100 Cummings Center
Suite 406-L
Beverly, MA 01915-6101
www.fairwindspress.com

15 14 13 12 11 1 2 3 4 5

ISBN: 978-1-59233-413-1

Digital edition published in 2012
eISBN: 978-1-61058-185-1

Library of Congress Cataloging-in-Publication Data

The country almanac of housekeeping techniques that save you money : folk wisdom for keeping your house clean, green, and homey / Richard Freudenberger and the editors of BackHome magazine.
p. cm.
Includes index.
ISBN-13: 978-1-59233-413-1
ISBN-10: 1-59233-413-X
1. House cleaning. 2. Housekeeping. 3. Home economics. I. Freudenberger, Richard. II. BackHome.
TX324.C67 2012
648--dc23
2011026276

Book layout by Daria Perreault
Illustrations by Judy Love

Printed and bound in Canada

THE COUNTRY ALMANAC OF HOUSEKEEPING TECHNIQUES THAT SAVE YOU MONEY

FOLK WISDOM FOR KEEPING YOUR HOUSE CLEAN, GREEN, AND HOMEY

RICHARD FREUDENBERGER
AND THE EDITORS OF
★*BACKHOME MAGAZINE*★

FAIR WINDS

CONTENTS

INTRODUCTION

Keeping house is an art, a science—and a tradition. But housekeeping today is getting more complex and more controversial despite technology that is designed to make life easier. As consumers, we confront daily decisions concerning the products we use (or avoid) in the home. Our store shelves are stocked with commercial cleaners that could pose serious health risks. Our grocery stores boast aisles of prepared foods designed to take the cooking out of mealtime. In general, the modern way of housekeeping is focused on instant gratification, less labor—just getting it done. The joy of creating a healthy hearth seems to be lost. We at *BackHome* magazine think it's time to return to our roots and keep life simple. It's time to revisit age-old, effective, low-cost ways to clean the home, prepare food, entertain friends and family, garden and preserve produce, celebrate holidays, and organize our homes and lives. We can save a significant amount of money by going back to the basics, and in turn, we'll reduce our environmental impact.

The concept of sustainability that has developed so much groundswell in recent years is not a brand-new idea. In the country, we have always found ways to maximize the useful life of our resources, reduce our reliance on energy, and become more self-reliant. And today, we continue to find ways to reuse, replenish, reinvent—to do more with less, to live richer by spending little.

Sound impossible? Actually, generations of country dwellers and our contributors at *BackHome* share just how easy it can be to live a self-reliant, budget-friendly lifestyle. The ultimate reward: greater independence! When you cut more clutter and cost out of life, you find room to really live.

This book is chock-full of how-to projects, recipes, and instructions for living a simpler, cleaner life. (Your wallet will thank you for it.) Our seasoned veterans provide practical, doable ideas you'll want to try. It's time to get back to tradition of keeping house, appreciate the art, and recognize proven, old-fashioned science that is safe, effective...and fun!

Now, it's time to get your hands dirty! (And we bet you'll find a new hobby in these pages while you're at it.) Let's get back to the basics and focus on what's important. That is the objective of this book. Now it's up to you to press "reset" in your own home and live a simpler, more fulfilling life. We're eager to help you on that journey.

Sincerely,
The editors and contributors of *BackHome*

PART 1
ALL AROUND THE HOUSE

CHAPTER 1

KITCHEN AND PANTRY

The kitchen is the heart of the home. It's the hub of activity, where we cook, dine, socialize, work—it's where everyone gathers. We want that area to be clean, well organized, and free of contamination and health risks. This chapter contains the basics: toxin-free cleaning, smart space-saving organization, and tips on tools and utensils. Here, you'll discover a world of money- (and energy-) saving ideas.

Avoid Toxins in Home Cleaners

Cost savings

About $25 saved by not having to purchase many cleaning, laundering, and toiletry items

Benefits

Improved indoor air quality

BackHome **contributor Shawn Dell Joyce brings in-depth focus to issues that connect us with our environment. Here, she hits home with a discussion of healthy, cost-effective household cleaner alternatives.**

Ever notice how most cleaning products do not have ingredient lists? These common products are not required by law to list what's inside, even though some might contain chemicals that can harm you and the environment. In fact, of the 80,000 chemical compounds currently used in commercial cleaning products, only one-half percent have been tested for cancer-causing ingredients, according to *Naturally Clean* author Jeffrey Hollender.

We breathe these chemicals while we clean, drink them in polluted water, and absorb them through our skin. (Kind of scary, isn't it?) The U.S. Environmental Protection Agency (EPA) estimates that we are exposed to air pollutants two to five times more indoors than outdoors. This is mainly because these pollutants are concentrated when we use them indoors with little ventilation.

To protect yourself and your indoor air quality, phase out household cleaning products containing any of the ingredients listed in Common Household Toxins on page 11.

Before we became so reliant on chemistry, we used to clean our homes with natural cleansers. Try restocking your cleaning cabinet with these natural and nontoxic alternatives:

- Pure vegetable or castile soap can replace shampoo, hand soap, and dish detergent.
- Vinegar cuts grease and removes mildew, is a brick/stone cleaner, and cleans windows without streaking. Drizzle it on top of a little baking soda to scour a toilet or clean a dirty bathtub.
- Baking soda (sodium bicarbonate) is a gentle abrasive powder for stovetops and shiny surfaces. It will extinguish a grease fire, deodorize your home, and can be used as toothpaste and deodorant.
- Borax is a mineral that can inhibit mildew, remove stains, work with soap as a laundry detergent, and be mixed with sugar to kill cockroaches.
- Cornstarch makes an excellent window cleaner or furniture polish when diluted with water. In its powder form, use it to shampoo carpets and starch clothes.
- Washing soda is a heavy-duty cleaner for burned pans, grills, and ovens.

Here are a few recipes with which to replace common household cleaners with inexpensive nontoxic alternatives.

- **All-purpose cleaner**: Dissolve 4 tablespoons (55 g) of baking soda in 1 quart (946 ml) of warm water.
- **Spray disinfectant**: Fill an empty spray bottle halfway with isopropyl alcohol and fill it up with water.
- **Drain cleaner**: Pour ½ cup of baking soda (110 g) down the drain followed by ½ cup (120 ml) of vinegar. Do not try this if you have already used a commercial drain opener, because the fumes would be toxic.
- **Window cleaner**: Refill a spray bottle with equal parts of vinegar and water. Add cornstarch for stubborn dirt, and use newspapers to clean the windows. Avoid cleaning windows on a sunny day so they won't streak.

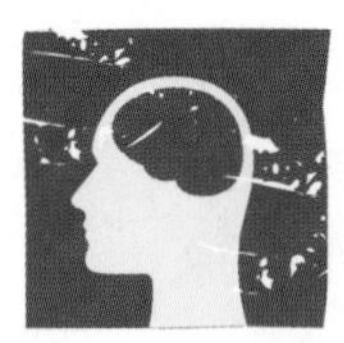

DID YOU KNOW?

The leading cause of in-home air pollution is (surprise!) the dishwasher. That convenient, labor-saving device is actually the most toxic appliance in your home. Researchers at the EPA and the University of Texas documented through 30 experiments that the super-heated water in dishwashers can cause chloroform, radioactive radon, and other pollutants to become airborne in your home. Dishwashers vent about 1½ gallons (6 L) of air per minute into your home during the wash cycle. They continuously release waterborne toxins into your air, sometimes in one huge burst of contaminated steam if you open the door too early.

- **Use chlorine-free, phosphate-free dishwashing soap.**
- **Ventilate your kitchen while the dishwasher is running by opening a window.**
- **Select the "no-heat dry" option on your dishwasher if it's available; if it's not, keep your dishwasher sealed shut for at least an hour after it has finished.**

Common Household Toxins

Clean your home and the environment by avoiding these common household toxins:

ETHOXYLATED NONYL PHENOLS (NPES) are classified by the Environmental Protection Agency as an "inert of toxicological concern." NPEs induce female characteristics in male fish, according to a 1996 study in the *Journal of Pesticide Reform*; and NPEs can cause an increase in breast cancer cells and reduce fertility in women. You find NPEs in pesticides such as Vinyzene and Nytek, and wood preservatives such as Socci and Cunilate. It may also be found in industrial cleaners and laundry detergents. Europe has banned products containing NPEs, but they are still in use in the United States.

METHYLENE CHLORIDE can be found in paint strippers and polyurethane foam. But it is most commonly used as a fumigant for strawberries, and most visible in those bubbling Christmas lights as bubbly red liquid. Methylene chloride is listed as a possible human carcinogen by the International Agency for Research on Cancer because it is metabolized in the body to form carbon monoxide.

NAPHTHALENE is often used in mothballs and moth crystals. Naphthalene is listed by California's Office of Environmental Health Hazards Assessment as a substance "known to the state to cause cancer."

SILICA is made from finely ground quartz, making it a carcinogenic breathing hazard. Silica is found in some abrasive cleansers and scrubbing powders.

TOLUENE is used as a solvent in numerous products, including paints. It is also sold as the pure product and is listed by California's Office of Environmental Health Hazard Assessment as a reproductive toxin that may cause harm to the developing fetus. Pregnant women should avoid products containing toluene.

TRISODIUM NITRILOTRIACETATE (NTA) is listed as a possible human carcinogen by the International Agency for Research on Cancer. It can sometimes be found in laundry detergents. NTA impedes the elimination of metals in wastewater treatment plants. This can cause metals that have already settled out to be brought back into the liquid waste stream.

XYLENE is another extremely toxic ingredient that can be found in graffiti and scuff removers, spray paints, and some adhesives. It is a suspected reproductive toxin that has shown reproductive harm in laboratory experiments, making it unsafe for pregnant women. It can also cause memory loss with repeated exposure.

BLEACH (SODIUM HYPOCHLORITE) is a common household cleaner that can be found in most commercial cleaning products. When bleach is mixed with a toilet bowl cleaner, it reacts to form deadly chlorine gas. When mixed with ammonia, it creates chloramine gas, which is toxic. In the marine environment, bleach bonds with organic material to form organochlorines, toxic compounds that can persist in the environment.

PHOSPHATES have been eliminated from most laundry products, but still persist in dishwashing detergents. Some soaps may contain 20 percent phosphates with high levels of bleach as well. Phosphates cause algae blooms that deprive aquatic species of the oxygen they need, choking rivers and lakes.

2-BUTOXYETHANOL, also known as **ETHYLENE GLYCOL BUTYL ETHER**, is used as a solvent in carpet and specialty cleaners. It can be inhaled when sprayed, or absorbed through the skin, and may cause blood disorders, as well as liver and kidney damage. According to the New Jersey Department of Health and Senior Services, it may also cause reproductive damage after long-term exposure.

Be a Kitchen Sink Alchemist and Save a Bundle

Cost savings

Between $3.99 and $18, depending on the product you're replacing

Benefits

Natural, nontoxic, and inexpensive substitutes for common household products

Chemicals that are usual suspects in today's commercial household cleaning products today didn't exist in "the olden days," and they got along just fine without them. The basic ingredients were mostly natural with no petroleum-derived synthetics or artificial additives—and they're still available today. So, let's take a look at how a little kitchen sink alchemy can still go a long way.

PAINT STRIPPER

This is an inexpensive and reasonably effective stripping solution for paint that uses washing soda (sodium carbonate) as the prime ingredient. Washing soda, also known as soda ash, is fairly alkaline, so protect your hands with rubber gloves when using it. The mixture is simple:

- Make a paste of the soda powder and slather it onto the surface to be treated with a rag or a broad putty knife.
- Keep the paste moist by spraying (not soaking) it over the course of 8 hours or so.
- Peel off the bubbled-up paint skin with a scraper. You do not want to leave the caustic soda in the wood grain, so it can be neutralized with a weak solution of white vinegar and water, mixed in a blend of 20 percent vinegar, 80 percent water.
- Allow the wood to dry completely after rinsing.

WOOD INSECT TREATMENT

Borax is a natural insect repellent and is blended with cellulose fibers to make an insulation that is insect and fire resistant. Boric acid, a constituent of borax, is an antiseptic powder with similar repellent qualities. Mix it with water to the consistency of a thin paste, and paint it on to wood surfaces to prevent bug infestation from household pests such as ants and roaches.

ASPHALT AND TAR REMOVER

Remove asphalt and tar spots from bicycles, car panels, shoe soles, and other surfaces with a mixture of white distilled vinegar and raw linseed oil.

- Place a few drops of linseed oil into 4 ounces (120 ml) of white vinegar and stir.
- Apply with a cloth and rub; then dry with a clean cloth.
- You can substitute glycerin for the linseed oil for cleaning.

LIME WHITEWASH

Whitewash historically was an inexpensive substitute for paint. It's still inexpensive and offers a nostalgic look to fences, outbuildings, walls, breezeways, signs, and even craft pieces. Its main constituent is mason's lime, the hydrated or slaked lime (calcium hydroxide) also known as Type S used by traditional brick masons. (The agricultural lime we use in the garden has not gone through the heating and soaking process to make it "slaked," and is not an acceptable substitute.) The beauty of whitewash is that it allows the wood beneath to breathe, plus it has natural insect-repellent qualities. Here are some pointers when using whitewash:

- Mason's lime comes in 50-pound (22.7 kg) bags, so unless you have a lot of area to cover and helpers to do it, you'll have to split the ingredients into more manageable lots.
- Fifty pounds (22.7 kg) of mason's lime requires about 6 gallons (22.7 liters) of clean water (preferably rainwater or soft well water, without chlorine) to bring it to a pasty consistency; mix it with a hoe in a tub or wheelbarrow.
- Apply the mixture with a broad stiff brush; it will dry leaving a rough texture. For a longer-lasting, more water-resistant coating, add 1½ cups (255 ml) of boiled linseed oil. For smaller lots, reduce the ingredients proportionally.

Save on Space with a Pantry Door Storage Unit

Cost savings

Between $18 and $27 in building rather than buying a storage shelf

Benefits

Gaining extra storage space in a convenient location

Kitchens and pantries are notoriously short of storage space no matter how big they are. But one handy storage spot that's usually wasted on bulky brooms and aprons is the inside surface of a pantry door. It's convenient, out of the way, and perfect for holding the jars and cans that tend to get lost in the back of larger cabinets.

You can construct a simple set of case shelves that you can fasten to the door to hold everything from canned goods and spice jars to plastic wrap and aluminum foil. It's nothing more than an 18 x 32-inch (46 x 81 cm) box, 5½ inches (14 cm) deep, with four shelves and stays at the front of each to keep the goods from tumbling out when the door is opened or closed. You can make the whole case from 1 x 6 shelf board.

MATERIALS

- 1 x 6 shelf board (5½ inches [14 cm] wide) — refer to cutting instructions steps 1 and 2
- (3) dowel rods
- (3) ⅜-inch (9 mm) dowel stays
- Wood glue
- Carpenter's saw (with dado blade)
- Drill and ⅜-inch (9 mm) bit
- Sandpaper (fine grit)
- Wood stain or polyurethane blended stain

STEPS

STEP 1: PREPARE THE WOOD PIECES:

1. From the shelf stock, cut two ¾ x 5½ x 32-inch (1.9 x 14 x 81 cm) boards
2. Also cut five ¾ x 5½ x 17½-inch (1.9 x 14 x 44 cm) boards

Beginning at one end of each 32-inch (81 cm) board, measure and mark line perpendicular to the long edges of the boards at these three points: 9½ inches, 19 inches, and 25½ inches (24, 48, and 65 cm).

4. Continue by marking a line 3/4 inch (1.9 cm) from each end of the boards
5. Measure and mark holes for the three dowel stays. These are 1 1/2 inches (3.8 cm) above the marked lines set back from the front edge by 3/4 inch (1.9 cm).

MAKE YOUR CUTS

1. Adjust the height of the saw blade to 1/4 inch (6 mm) and make the four end rabbit cuts at the marked points. It's best if you use a dado blade to remove the wood full-width all at once, but you can make a series of cuts with a single blade if necessary.
2. Go back and finish the four dado cuts on the sides, in each case cutting to the short side of the mark so as not to remove too much wood.
3. Use a 3/8-inch (9 mm) bit to drill the dowel holes at the six marked points. These should do not penetrate the sides, but should stop at a depth of 3/8 inch (9 mm).
4. Sand the wooden components lightly, then cut three dowels to 17 inches (43 cm) in length.

ASSEMBLE THE CASE

1. Spread a layer of wood glue onto the edges of the top and bottom boards and the three shelves.

Fit the five pieces into their respective slots in the side boards. At the same time, you'll need to slip the ends of the 3/8-inch (9 mm) dowels, which are not glued, into their sockets. (It would help to enlist some assistance at this point.) Set the case on a flat work surface and the corners checked for square before clamping the case for drying.

FINISH THE SHELF UNIT

1. After the glue has cured (12 to 24 hours), tack or staple the back panel in place after centering it over the opening.
2. Sand the entire case with a fine grit paper, rounding off the sharp facing edges of each board.
3. Finish inside and out with wood stain or polyurethane blended stain. The unit can be fastened to the inside of the door with a few wood screws at the top and bottom of the case.

Make Your Own Nontoxic Odd-Job Kitchen Cleaners

Cost savings

Between $2.40 and $12, depending on the product

Benefits

Safe and nontoxic cleaning for the kitchen

Even clean kitchens get dirty fast, maybe because they see a lot of traffic. Still, when it comes to sprucing up, the kitchen is a showcase of a lot of different materials—metals, porcelain, wood, Formica, tile, perhaps some marble or soapstone, or maybe acrylic polymer compounds such as Corian. Each has its own requirements when it comes time to clean, but once again, common household ingredients come to the rescue.

COUNTER TILE

Clean porcelain tile with a half-and-half mixture of white distilled vinegar and water. Spray the mixture from a bottle or wipe it on the surface with a cotton cloth. To remove grime and stains from grout, make a paste from baking soda and a few drops of water and scrub grout with a toothbrush. Borax can be substituted for baking soda for tough jobs.

FORMICA

Formica surfaces and countertops respond well to a blend of baking soda and castile soap, or a mild dish soap. You don't need to make too much at one time; 1/4 cup (55 g) of baking soda mixed with enough soap to make a light paste will go a long way.

PORCELAIN ENAMEL

This treatment works well on porcelain enamel, found on appliance surfaces, sink fixtures, and cookware. White distilled vinegar is antifungal and a disinfectant, and will brighten up dulled porcelain. Mix 1 part vinegar to 4 parts water, and wipe it on with a clean cloth. Allow it to work for a few minutes, then wipe it off with a wet cloth.

KITCHEN WOOD

Clean grease from cabinets with a lemon juice/castile soap formula. Mix 1/3 cup (+80 ml) of lemon juice with 1/4 cup (+60 ml) of water, and add 1/2 teaspoon (2.5 ml) of liquid soap. A small amount of extra-virgin olive oil can be added for lubrication. Apply with a damp sponge and let dry (do not wipe off). The lemon leaves a pleasing scent.

ACRYLIC POLYMERS

Use a gentle cleaner on countertops made of Corian and similar manufactured solid surfaces (except for products made by DuPont). Mix a few drops of liquid dish soap, 1 tablespoon (15 ml) of white vinegar or lemon juice, and 1 pint (473 ml) of warm water to make a mild cleaning solution that can be safely wiped on to get rid of soil and everyday stains. Do not wipe off.

SOAPSTONE

Soapstone, or steatite, is treated with mineral oil to give the surface a pleasing dark patina. Avoid using strong chemicals on these counter surfaces, including household ammonia. Instead, use a mild liquid soap (or a scented castile soap) and water. Do not use acetic solutions such as white vinegar or lemon juice, as they can corrode the stone particles.

MARBLE

Marble boards and surfaces are harder than soapstone, so a mild cleaner such as borax can be used without damaging the surface. Moisten a soft cotton cloth with warm water and dip it into the powder, then rub it on. Wipe with warm water afterward. Never put vinegar or lemon juice treatments on marble; they will react with the calcium carbonate and can etch the surface.

ALUMINUM, STAINLESS, AND COPPER

For safe and simple cleaning, blend of 1/4 cup (60 ml) of white distilled vinegar and 3/4 cup (175 ml) of water. The mixture is a perfect disinfectant and degreasing solution for most kitchen metals, even copper. Avoid baking soda because it can pit aluminum.

VINYL FLOORING

Vinyl is easy to clean because it is resistant to so much. In 1 gallon (3.79 L) of warm water, mix 1 ounce (28 ml) of castile soap and 1 ounce (28 ml) of white distilled vinegar. Choose a scented castile soap for a pleasant scent (the vinegar odor will subside). Use a sponge mop to wipe the floor surface, and allow it to air-dry.

Clean Your Appliances Naturally

Cost savings

About $1 to $5.50 per treatment, depending on the appliance

Benefits

Natural and nontoxic, chemical-free cleaning

Heat, steam, cooking odors, and perishable foods are bound to leave their mark on the kitchen, and especially on appliances. These tips for cleaning appliances will help your wallet and the environment.

STOVETOPS AND OVENS

Range burner pans and stovetops are often the victims of spills and boilovers. It's usually not practical to deal with such messes right away, but as soon as possible, wipe up the spill with a damp sponge dipped in baking soda. For sauces and oily spills, using sprinkled cornstarch in lieu of baking soda works well because of the absorbent qualities of cornstarch. The soiled paste can be wiped away easily once it has soaked up the spill.

Clean oven racks and drip pans outside the appliance, in the sink or a large tub. Baking soda and table salt, along with soap and water, work well on enameled pieces; the racks may need a steel pad if they are particularly soiled. Clean the oven interior with a layer of baking soda powder moistened with a water spray. Leave the paste on these surfaces for 8 hours or more, then wipe away with a moist cloth, using a plastic scraper if needed. Rinse thoroughly before using the oven. For stubborn stains, blend a mixture of 1 part borax with 2 parts baking soda—the alkalinity will help lift soil.

TOASTER OVENS AND TOASTERS

These small appliances have more exposed circuitry than other types of kitchen equipment, so it's important they not be doused with liquid when cleaning. Always unplug your toaster before cleaning it. With oven-style toasters, remove the racks and steel pan(s) and clean them in the sink using soap and water. For more stubborn stains, sprinkle baking soda on a cleaning sponge. There may be a cleanout door at the bottom, which you can open over the sink to remove crumbs. With conventional toasters, turn the apparatus upside down and shake out the debris. Clean the outside of the cabinet with soap and water, or for tougher stains, mix 1 part white distilled vinegar with 1 part water. The glass door on the oven will shine up with vinegar as well.

REFRIGERATORS

A large part of effective refrigerator operation is in removing moisture, which aids in efficient cooling. In order to remove moisture, air must circulate through the interior, which provides a perfect opportunity for the odors of stale and spoiled foods to permeate the chamber. It is easiest to clean a refrigerator when there's no food in it, so you should have a cooler ready to hold the contents while you're cleaning. Cleaning doesn't take much—a sponge and 1 gallon (about 3.8 L) of warm soapy water with 1 cup (about 220 g) of baking soda will do the trick. (If there are any aluminum parts inside, forgo the baking soda or the metal may pit. Also avoid soaking interior lighting fixtures and controls.)

Use this formula for the entire refrigerator interior, including shelves and bins. The same soap and water can be used on the outside cabinet and the handle. Sticky handles will come clean with a white vinegar/water solution.

COFFEEMAKERS

Store-bought products marketed to clean coffeemakers are generally acidic and toxic if ingested. The same goes for the bathroom rust and lime removers labeled for use as coffeepot cleaners. In most cases, straight white vinegar does an excellent job of removing scale and stains from the internal surfaces and parts of coffeemakers; the only requirement is that it be given time to work, generally a few hours. Percolators and automatic drip machines can be run through a cycle with white vinegar to clean them. Try using a bottle brush to scrub stubborn stains. Rinse with water after cleaning.

DISHWASHERS

Most problems with automatic dishwashers occur from residual odors. The easiest way to resolve this problem is to run an unloaded cycle with 1 cup (about 220 g) of baking soda or borax (no soap). It will freshen the tub and racks and soften the cleaning water so the machine will work more effectively.

MICROWAVES

Mild soap and water is all you need to clean the outside cabinet of a microwave. Mix 1 part white vinegar to 3 parts water to clean the door. Take out the glass or plastic food tray at the bottom and clean it separately in the sink with soap and water. The tougher cleaning problems arise inside, where splattered foods get cooked onto side and top panels. To preclean, place a bowl of water in the microwave and set on high for 3 to 5 minutes, until steam forms. This will loosen up splattered food stains. For tougher spots, use a mild soap and baking soda paste, using a plastic scraper if necessary.

BLENDERS AND FOOD PROCESSORS

The beauty of these small appliances is that they can often be cleaned effectively simply by adding soapy water and turning them on for a few moments. If individual blades or paddles need cleaning, carefully remove them from the tub (if they are removable; in blenders they may not be) and try a vinegar wash followed by water rinse. Allow parts to dry before storing the machine.

Simple Solutions for Shining Silver

Cost savings

Silver polish is fairly inexpensive, costing less than $10, but you can pocket that money by using products you already have at home to clean silver.

Benefits

A safe alternative to commercial silver polish that effectively removes tarnish.

Rather than rubbing fine silver with noxious polish, shine up your special utensils with a household basic: toothpaste. Simply dab some paste on your finger or use a clean cloth, and rub into tarnish.

Polish platters and larger silver pieces with baking soda and a clean, damp sponge. Make a paste of baking soda and water, and rub the concoction in to the silver. (It is nonabrasive but will work away tarnish.) Rinse with hot water and polish with a cloth.

Make Your Own All-Purpose Surface Wipes

Cost savings

up to $10/month on wipes; or up to $120 per year, depending on frequency of use

Benefits

Clean surfaces with all-natural solutions and reduce waste with reusable cloths.

The ease of commercial cleaning wipes is hard to beat—but the price of disposable wipes designed to swipe away kitchen and bath grime can really clean out your wallet. Considering how many of these wipes you might use in a week, you can save money and the environment by making your own nontoxic, reusable wipes that do the same job. You'll reduce waste and ensure that the solutions that touch your home's surfaces are safe. (Doesn't it make sense to use edible substances on surfaces you use for eating and cooking?)

Once you get in the habit of making your own cleansing wipes, the process will take you less time (while using no fuel) than hauling to the store to buy more cleaning wipes.

MATERIALS

- Container (plastic container with a "pop" lid; or reuse a cleaning wipes container)
- A package (or more) of reusable cloth wipes
- Distilled white vinegar
- Water

STEPS

- **Mix the cleaning solution.** You can adapt this recipe to make cleansers for various purposes (such as glass cleaner). This mixture is an all-purpose surface cleanser using a 1:2 mix of vinegar and water.

- **Prepare the cloths.** Fold and soak the reusable cloths in the vinegar-water mixture.
- **Package the wipes.** Keep the wipes moist by placing them in a container. A plastic (BPA-free) container with a pop-out top provides easy access. Or simply use a saved container from the disposable wipes that you no longer have to buy now that you can make your own!

Freshen Garbage Pails and Disposals

Cost savings

Minimal, but you can rely on ingredients you already stock in your kitchen.

Benefits

Prevent kitchen odor and bacteria growth in trash containers; and avoid using chemical cleaners in the cooking environment.

Prevent garbage pail stench with a quick and easy deodorizing solution that can be adapted to freshen a garbage disposal. All you need is baking soda, salt, mild dishwasher detergent, and water. (If you have extra citrus peels, save those, too.)

SWEETEN REFUSE CONTAINERS. Every time you empty your garbage container, sprinkle some baking soda in the bottom of the pail. Periodically wash and deodorize the can with baking soda and warm water.

DE-GUNK THE DISPOSAL. Pour 1/4 cup each of salt, baking soda, and dishwasher detergent into the disposal. Flush with hot water while running the disposal. This mixture will remove buildup and neutralize odors. If you have citrus peels on hand, drop the rind down the disposal and turn it on while flushing with water. A fresh scent will linger, and the rind helps clean the "teeth" of the disposal.

Make Your Own Hand Sanitizer

Cost savings

You'll spend about equal, but win by making a healthy alternative to drying commercial sanitizers.

Benefits

This moisturizing hand-sanitizer contains no alcohol like off-the-shelf varieties.

Pocket-size gel hand sanitizers are handy for on-the-go "cleanup," but these formulas are drying, costly, and contain 60 to 90 percent alcohol. Why rub a formula on your hands that is potentially toxic before you eat? The risk elevates for mothers who carry sanitizers to clean children's hands. The good news: You can make your own safe, affordable hand sanitizer.

Witch hazel, tea tree oil, and peppermint have antiseptic and astringent properties. They're great germ killers. Aloe vera will prevent the skin from drying out—a welcome alternative to commercial sanitizers that leave hands parched.

Note: Even use of this all-natural solution should be limited to a few times daily for those with sensitive skin.

MATERIALS

- 1 cup (235 ml) pure aloe vera gel
- 1 to 2 teaspoons (15 to 28 ml) witch hazel
- 3 to 4 drops tea tree essential oil
- 3 to 4 drops peppermint essential oil

STEPS

STEP 1: Stir together ingredients.
Measure the ingredients and stir together in a mixing bowl.

STEP 2: Adjust aloe vera content.
To thicken the solution, add more aloe vera gel one teaspoon at a time.

STEP 3: Package the product.
Travel-size toiletry bottles are ideal for toting this hand sanitizer. Save and clean empty bottles for reuse if you don't want to invest in empty new ones (available at most drugstores).

Keep Pests Out of the Kitchen

Cost savings

The cost of any potential future pest control

Benefits

Handle pest control before it becomes necessary without harmful chemicals or dangerous traps.

Watching a roach travel across your kitchen floor, or a mouse scurry from the pantry where you store food will certainly make you lose your appetite. Who knows how many other pesky relatives are renting space in your cupboards—once you spot a critter, there are surely more of its kind close by. But how do you handle pests in a place where food is stored, prepared, and served? If we focus on using cleaning products that are virtually edible, the same care should be applied to how we treat pests.

You can control pests and prevent infestations while using an environmentally friendly tactic known in the pest control industry as integrated pest management (IPM). IPM is a sensitive approach to pest management that relies on a combination of common-sense practices. It involves managing pest damage by the most economical means using strategies that cause the least possible harm to humans and the environment.

Specifically, IPM involves these steps, and you can apply these methods to your kitchen pest-control approach.

Decide when to take action. A single pest sighting does not necessarily mean your kitchen is under siege by pests (spiders, ants, roaches, mice, you name it). Decide at what point the pests are a real threat. Then take action.

Monitor and identify pests. Assuming you're not an entomologist, this part of IPM may require professional insight. But you can do IPM yourself by recognizing what pests need control, and which are harmless to your kitchen environment. (Some might argue that all pests are harmful in the kitchen—but that depends on the limits you set.)

Prevent pests from coming inside. IPM calls for managing outdoor spaces so pests do not become a threat indoors.

Control is your last resort. If pest control action is required, effective less risky pest controls are chosen first using targeted methods. In other words, don't blanket treat your entire kitchen: Identify the problem spot, and treat that spot. (For specific pest control recipes, see page 111.)

IPM AT HOME

Translate IPM to the household—and specifically, the kitchen—by taking some simple steps to keep pests out of your cooking and food storage environment.

- Store food in sealed containers.
- Regularly remove garbage from the kitchen.
- Place garbage containing food scraps in tightly covered trash cans.
- Fix leaky plumbing.
- Do not allow water to accumulate on surfaces (floors, countertops, corners).
- Put away pet food at night.
- Control clutter: keep shelves neat and clean, and wipe them regularly.
- Caulk cracks and crevices around cabinets and baseboards.

Take a Fresh Look at Cast-Iron Cookware

Cost savings

Over a 40-year lifespan, cast-iron cookware can save about $500 in replacement costs over nonstick cookware

Benefits

Energy and water savings, healthy cooking

Not so long ago, cooks all over the landscape were abandoning their cast iron in favor of lighter, sleeker, more colorful cookware choices in the aluminum line—especially the slick-skinned Teflon-coated beauties that bypass the need for seasoning.

But that old stalwart, cast iron, seems to be getting a new lease on life these days. For one, preparing meals in cast iron is a healthy way to cook. Even food cooked in a well-seasoned iron pot or pan will absorb some degree of dietary iron, which is a necessary nutrient, especially for groups at higher risk of iron deficiency, such as children and women of childbearing age.

Cast iron is also energy-efficient if used properly, and, as it turns out, it is also more resource efficient than its counterparts. "Resource efficient" may sound like gibberish, but how many nonstick pans have you thrown away once the magic coating has scratched or worn through? (Using pans with damaged nonstick coating is not healthy.) Buying and replacing otherwise good pans every few years is environmentally irresponsible.

And speaking of resources, how much water do you use cleaning those nonstick pans? They're not dishwasher safe, so they must be washed, soaked, and rinsed by hand. What's more, they can't be safely scrubbed with steel or nylon pads, so the universal solvent—water—is usually the solution we look to.

What about energy use? Well, that depends on how you cook, but the average person is likely to overheat cookware, believing that cooking faster equals cooking better. For some types of cooking—the thin-skinned wok comes to mind—this is okay because the heating period is brief and the vessel's thin steel transfers heat effectively. However, cranking up the electric element or allowing a gas flame to lick out beyond the perimeter of a pan can not only damage a modern pot or pan, but it also wastes heat.

Cast iron, on the other hand, is a natural heat storage medium, and flame levels can be kept moderate (read energy-saving) and still get the cooking done. The wonderfully even heat dissipation imparted by a cast-iron Dutch oven or lidded pot is second to none, and with slower cooking times, those roasts, stews, and casseroles just simmer on low heat, with excellent through-cooking and very little moisture loss. (Searing meat requires higher heat, of course, but only for a brief period.) .

Care and maintenance of cast iron is almost legendary.

SEASONING IRONWARE

Wash new ironware in warm water with dish soap and dried with an old towel that you don't mind staining. After drying completely, coat it inside and out with lard or unsalted vegetable shortening (do not use butter or margarine as they may contain salt). Place the cookware bottom up onto the top rack of a 300°F (150°C or gas mark 2) oven and put a cookie sheet or layer of foil beneath it to catch the drippings. Bake the ironware for 30 minutes, then turn off the heat and let the pan cool slowly in the oven until you can handle it comfortably. Wipe off any excess oil and store the pan in a dry location. Any lids and iron accessories should be seasoned in the same manner and kept dry as well.

It's always a good idea to cook higher-fat foods or fry something like potatoes in oil the first few times after seasoning. Once the piece has been "broken in," you can use it normally.

CLEANING IRONWARE

Wash the pan immediately after use, while it's still warm, and allow it to dry thoroughly (a warm stovetop is perfect for this). There are a few "don'ts" with cleaning ironware:

- Never allow food or liquid sit in cast iron, and do not store food in iron vessels. High-acid foods will leach iron and eventually can pit the surfaces of the pan.
- Never pour cold water onto a hot iron vessel (it may crack).
- Never wash iron in a dishwasher or with detergent unless you plan to re-season it immediately.
- Be cautious when scraping out sticking food residue because you may remove the seasoning. Note, however, that scraping is far better than soaking, which shouldn't be done.
- Properly seasoned and maintained, cast iron will more than rival your best nonstick cookware, and you'll feel better using it as well.

Wok Your Way to Energy Savings

Cost savings

Several cents per meal in energy costs (it adds up!)

Benefits

Energy-saving, quick cooking that maintains foods' nutritional value

Unless you're already an Asian cooking buff, you may not realize what a valuable addition a traditional wok makes to your kitchen inventory. Despite its unwieldy size and peculiar bowl-like shape, the wok has several things going for it—things that you can get going for you if you take note.

FAST FOOD PREPARATION

Chinese cooks developed the thin-walled steel utensils so they'd heat up and cook very rapidly at high temperatures, thus conserving heat and energy. As a bonus, foods cooked this way lose less of their natural juices and absorb less oil in the process. Consider how a wok meal is cooked in a matter of moments. Over a month or year of cooking, the energy factor adds up.

CONSTRUCTED FOR EFFICIENCY

The traditional wok has a rounded bottom; departures from tradition only occurred when partly flat-bottomed versions were made to suit the heating elements of electric stoves. Either design is fine, but if you're splitting hairs, stick with the round bottom and ring stand for gas cooking and flat-bottom designs for electric cooking, because heat will transfer more efficiently to the vessel if you do.

A traditional wok is made of thin carbon steel, which absorbs and transfers heat quickly. Ideally, the metal is thin—so thin that fussy Chinese citizens have been known to take their newly purchased woks home and hammer-forge them under flame to thin the metal more to their satisfaction. Woks are also available in cast iron and with nonstick coatings, which require special cooking utensils.

SIZED TO SAVE

In keeping with the wok's energy-saving function, foods are cooked in the bottom of the pan and, when done, they are pushed up the sides—where temperatures are not as intense—to make room for the next ingredients to be cooked. This also provides the benefit of not having to use more than one cooking utensil if you've prepared well.

USING YOUR WOK

Woks are usually associated with stir-frying, but they can also be used for sautéing, stewing, steaming, or deep-frying, so it's always a good idea to purchase a lid or cover for your wok when buying. One culinary habit you'll have to adopt when stir-frying is to prepare everything beforehand, as there is no time for prepping once the cooking has commenced.

Typically, when stir-frying, a small amount of oil—usually less than a tablespoon—is placed in the bottom of the pan and spread to the sides with a wok ladle or round-tipped spatula. Oils with a high smoke point that won't break down easily under heat and turn bitter are chosen for cooking. These might include canola, olive, grapeseed, and peanut (if no one is allergic). The sesame oil often associated with Asian food may not be the best choice for cooking; it is usually added as a seasoning after the fact.

CLEANING A WOK

Because traditional woks are made of carbon steel, which is more prone to oxidize than contemporary cookware, the cleaning technique is different than what you might be used to. After the cooked food is removed from the vessel, reheat the wok while adding water. Bring it to a simmer. Scrape food residue and pour out water. To dry, wipe with a paper towel, then lightly rub canola or peanut oil into the cooking surface with a cloth before storing the wok.

Try a Wood Cookstove

Cost savings

Between $3 and $10 per month, depending on how often conventional range and oven is used

Benefits

Reduced fuel consumption, additional heat in winter, classic looks

When it comes to cooking, *BackHome* contributor Mrs. John Bayles is no stranger to advanced technology. She has used a variety of electric and gas ranges. But she said that the most superior cooking stove she's ever used is in her kitchen right now: a 1932-model wood-burning cookstove.

You would think that a wood-burning cookstove is a step backward in technology. How could it possibly compete with a modern gas or electric range? Actually, the efficiency and convenience of an old-fashioned cookstove might surprise you. It saves fuel by heating the oven and range top at the same time, and you can fuel it with paper-trash and twigs. Most of all, these stoves are real charmers. Why would anyone settle for a porcelain box in the kitchen when the curvaceous personality of a wood-burner is available? I own three: a mint green porcelain beauty with elegantly curved legs, a small burner-only model, and my prized possession, a 1912 Globe—black cast iron with chrome trim, complete with overhead warming oven and a water warmer.

FINDING AND INSTALLING A WOOD COOKSTOVE

A brand-new wood cookstove can be expensive, but it's virtually maintenance free because these stoves are built to last. New cookstoves start at $500 and can cost as much as $5,000. Good mail-order sources are Lehman's and Antique Stoves. (See "Resources," page 26) Some newer models are completely disguised as modern stoves by running the stovepipe out the wall directly behind the stove. Personally, I like my pipe to stand tall and proud, and the high-heat paint comes in so many colors that the pipe can be turned into a work of art that matches your kitchen decor.

If you're willing to hunt at auctions and antique stores for treasures like these, you will pay much less. I found my Globe in a neighbor's barn (it cost $300). My friend sold the 1932 model to me for just $25.

COOKING CONVENIENCES

A wood-burning stove can act as a microwave, warming drawer, and steam oven (how convenient!). I used to think I could never live without a microwave, but I basically used it as a way to make instant coffee in 2 minutes. Then I found a way to offer a visitor a steaming, "freshly brewed" cup of real coffee in 30 seconds. With my present stove, I put on a pot of coffee every time I cook, then pour it into a thermos, which we've affectionately dubbed "the microwave." We gave away the original.

Also, breakfast at our home is a snap: Kindle up the firebox under the burners, and then stick the biscuits in the oven. By the time the bacon and eggs are ready, the biscuits are browned to perfection. And that smell of coffee and wood smoke early in the morning—this is living!

When I need the effect of a traditional oven, I use my top burners to steam food—anything from casseroles to desserts and homemade breads. Bread and pie crust don't get crusty when steamed, but burning food is nearly impossible because food cooks so slowly and stays moist. (It simply gets browner the longer it cooks.)

The only drawback to woodburner cooking is keeping the fire going for meals that require long-term cooking. For example, with a large turkey, you have to stick close to the oven to stoke it because the firebox under the oven box isn't very large. But if I use hedge wood, or osage orange, as fuel for long-term oven cooking, the wood burns long and hot and I don't have to stoke the fire nearly as often.

STEAM-OVEN COOKING

Convert any large pot with a tight-fitting lid (such as a canning pot) into a steam oven to make main courses, soups, desserts, and breads. Here's how to rig the pot to steam your food:

- Cover the bottom of the pot with a few inches of water, then put in some custard cups on which you can rest an inverted pie pan.
- Place whatever you are cooking into the pot and cover it with a tight lid or aluminum foil.
- Heat up the stove, listen for the water inside the steam oven to boil, and adjust the heat to maintain an even boil. (Though a really hot fire doesn't hurt anything; it just speeds up the cooking time.)

SUGGESTED COOKING TIMES:

Bread:	2 hours
Casseroles:	1½ hours
Pies:	1 hour
Vegetables:	30 to 45 minutes
5 pounds (2.3 k) of meat:	3 hours

*Check the food often, add water when needed, and adjust the times according to your own tastes.

Resources

Lehman's
P.O. Box 270
Kidron, OH 44636
888-438-5346
www.lehmans.com

Antique Stoves
410 Fleming Rd.
Tekonsha, MI 49092
www.antiquestoves.com

Dry Your Own Herbs and Keep Your Money at Home

Cost savings

Home-dried herbs cost less than one-tenth the price of packaged herbs.

Benefits

You control the selection, quality, and most importantly the freshness of the herbs.

Even in the dead of winter, you can enjoy your bounty of fresh herbs. Dry herbs in the fall and they'll maintain their essence for up to one year. The best part: You'll save a bushel on what you would have spent on store-bought herbs. Here, *BackHome* contributor Kris Wetherbee lays out techniques for drying common herbs.

Many kitchen gardeners hang bouquets of herbs in an arid, dark spot to dry. This is effective but, depending on what herbs you're drying and how fast you want to get them dehydrated and stored, there may be a better way. You can dry herbs in the oven, microwave, a commercial food dehydrator, or in a dry, airy closet.

HARVESTING HERBS

The best time to harvest herbs is at their peak, just before they flower. Pick herbs in early morning, after the dew has dried and before the sun begins to bake out oils that give herbs their flavor. Dry, overcast days are ideal. Choose only plants that are free from disease and insects. If you're collecting several different herbs, use a separate basket for each. Use scissors to cut the stems, and handle the herbs as little as possible.

Remove dead or wilted leaves and weeds. Leaves should be left on the stem unless you're drying a large-leaved herb such as lovage or borage.

Dry the leaves thoroughly. Moisture can result in a moldy finished product and could lengthen the drying process. Once they are cut, dry them thoroughly.

Wipe root spices clean. Be sure to dry them well when you're finished. Garlic, ginger, and horseradish should then be peeled and sliced thinly.

HANGING IN BUNCHES

Hanging in bunches works best for herbs with long stems: oregano, sage, rosemary, mint, lavender, and lemon verbena. Hanging herbs are a pretty sight in a kitchen, but this isn't the best place to dry them because of light and humidity. Instead, find a dark, dry, warm environment with good circulation. A barn, shed, or airy closet are all possibilities. Even a car parked in light shade with a thermometer inside will do. Ideally, the temperature should be about 80°F to 90°F (about 27°C to 32°C).

Bunches should be no thicker than 1 inch (2.54 cm) in diameter at the stem ends. Tie them with string, twist ties, or rubber bands. Leave plenty of room between bunches to allow for good air circulation. (I use a folding wooden clothes rack for hanging my herbs.) A fan set on low can also help. Drying time depends on the type of herb, weather, and humidity, but usually ranges from 5 days to 2 weeks.

If the location is dusty, punch a hole in the bottom of a brown paper lunch sack, pull the stems through, then hang the bunch. Make sure to cut a few slits in the sides to provide ample circulation. The bag will keep the herbs clean and free from dust.

Also use a paper bag when drying seed heads such as celery and dill, but don't punch a hole in the bag; place the bouquet in the sack so the stems come out the opening. Tie with a string, cut small slits in the sides of the bag, and hang by the stems. Any loose seeds that fall out will collect in the bag.

SCREEN DRYING

Short-stemmed herbs such as thyme can be difficult to hang-dry, but screen drying is a great alternative. This method is useful for flower heads and petals that are too delicate for the heat of a dehydrator, oven, or microwave. Also, comfrey and borage, with large leaves, sometimes grow in big bunches that retain water if you hang them but will dry well on a screen.

Begin with a clean window screen or a rack covered with cheesecloth, muslin, netting, or paper towels. Spread herbs in a

single layer with plenty of space between them. Choose the same sort of warm, dry environment as you would if you were drying in bunches. Set your screens on blocks to allow circulation underneath, and remember to stir the herbs occasionally so air reaches all sides. Drying time will vary from 7 to 10 days.

OVEN DRYING

Oven drying is useful if you want quick results without investing in a dehydrator. Also, items that take a long time to dry, such as roots and hot peppers, do well in the oven.

The oven dries herbs very quickly, but some will take longer than others, so don't mix different types. Rosemary may take only a few hours, whereas large-leaved herb and root spices can take up to 8 hours.

Keep the oven on the very lowest setting possible with the door ajar. Or if you have a gas stove with a pilot light, the flame will provide enough heat for drying herbs. In this case, leave the oven door closed, but open it occasionally to let out moisture.

Baking sheets work, but a cake rack that will allow the warm air to circulate evenly is better.

FOOD DEHYDRATOR

Gardeners who dry their fruit use commercial food dehydrators—small electric machines that use a heating element and fan to control the drying environment. Food dehydrators are great for drying herbs, too. Just about any herb works well with this superfast method of drying, including sliced roots such as garlic and ginger. My personal favorite for the dehydrator is celery leaf, a must for wintertime soups and stews.

During summer and fall, my dehydrator is in perpetual motion, drying every herb imaginable, plus plenty of Asian pears. I just load up the circular tray and plug in the cord. Nothing could be easier. Dehydrators work fast—thyme, for instance, might take 5 to 7 days to screen-dry but only about 8 hours in a dehydrator.

Here's how to prepare herbs for the dehydrator. Place a single layer of herbs on each tray of the dehydrator. You can dry different types of herbs at the same time, but only one type of herb per tray. If your dehydrator has a temperature control, set it between 90° and 100°F (about 32° and 38°C). If it doesn't have a control, regulate the heat by adjusting the vents.

MICROWAVE DRYING

The microwave dries herbs very quickly, making it useful for thick roots and herbs that are very damp or were rinsed after harvesting. You can use the microwave to start herbs that you plan to hang- or screen-dry. A minute in the microwave will take days off the drying process and insure against mold.

Pat herbs dry and place in a single layer between two paper towels. Microwave for 2 to 3 minutes, checking every 30 seconds. Microwaves vary dramatically in terms of power, and just a few too many seconds can result in scorched herbs, while an extra minute can incinerate your harvest.

STORING DRIED HERBS

To test whether an herb is dry, crumble a leaf between your thumb and index finger. If the leaf turns into a coarse dust, the herb is dried. If it wrinkles or shows any signs of moisture, you need to dry it longer. If you're drying roots, test them by bending them. If they snap, they are dry.

Strip leaves from stems, but do not crush them. Leaves and seeds left whole will retain their flavor longer. Grind roots in a spice grinder.

Store dried herbs in a tightly sealed container. Canning jars with rubber seals, airtight tins, or glass jars with screw-on lids are ideal. Recycled glass mayonnaise, mustard, and peanut butter jars all work well. For smaller amounts, try using baby food jars. Do not use plastic containers. Fill each jar loosely, then label and date it. Once dried, many herbs look the same and can be hard to tell apart.

Check the jar after a few days for any sign of dampness or moisture collected on the inside of the glass. If there is moisture present, remove the herbs and dry, dry again.

Keep containers away from bright light and any heat source. When adding herbs to a recipe, go away from the stove to measure the amount you need. Standing over a warm stove with an open jar is a quick way to add humidity and a sure way to sabotage your efforts.

By the time your supply of dried herbs starts to wane, your fresh herbs should be ready to harvest again.

Learn the Art of Fireless Cooking

Cost savings

Between $2.50 and $4.50 per month, based on how often a conventional oven is used

Benefits

Fuel conservation, convenience, a cooler summer kitchen.

Fireless cooking in a container is designed to maintain a steady temperature. It's efficient, cost effective, and convenient. How would you like to reduce cooking fuel use by 80 to 90 percent and have ready-to-eat meals for picnics, potlucks, or family dinners? Even better, food retains nutrients without overcooking and stays moist.

Fireless cooking is not a new concept. For centuries, the method has been used by campers, vacationers, and fisherman, and homemakers were encouraged to do the same with help from commercially made cookers and cookbooks tailored to the subject. Before that, the pit-cooking Dutch oven was in regular use, and who knows how long Hawaiian luau cooks have been practicing their skills.

***BackHome* contributor Greg Lynch is a big fan of fireless cooking and summarizes it here. It's simple to make your own fireless cooker. Here's how.**

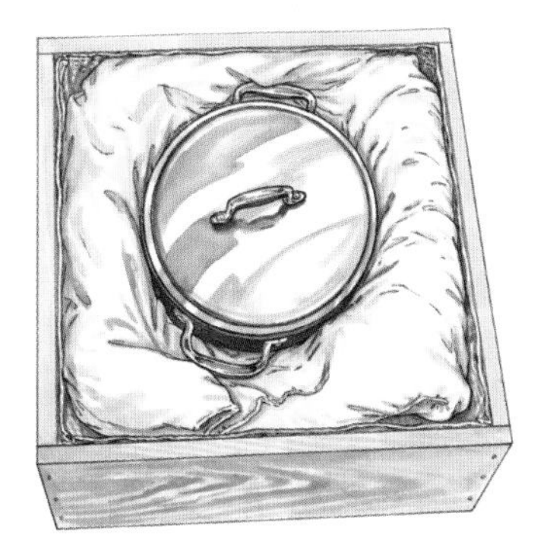

MAKE A FIRELESS COOKER

MATERIALS

- Seasoned, cast-iron pot with a lid or enamel-coated cast iron (stainless steel also works)
- An insulated chamber (such as a milkbox, heavy cardboard, plywood box, foam steak shipper, or a 5-gallon (18.9 L) pail are all possibilities)
- Radiant barrier (such as aluminum foil; or Reflectix or Astro-Foil, available at building supply outlets)
- 2-inch (5 cm) rigid polystyrene foam (used in housing construction)
- Glue

STEPS

Construct the Cooker

1. Choose a container that you'll fill until it's quite full while cooking. Full pots minimize heat loss, and larger pots maintain heat for longer. The mass of iron holds a great deal of thermal energy, which contributes to the heat available when it's placed in the insulated chamber.

2. Find an insulated chamber that is 4 inches (10 cm) larger than your pot on all four sides, the top, and the bottom.
3. Add a layer of insulation between the pot and chamber. This layer should be a radiant barrier, using aluminum foil at the very least. You can also use high-efficiency radiant barrier such as Reflextix or Astro-Foil. The barrier will reflect back infrared energy that the pot emits.
4. Create an outer polystyrene layer to wrap around the insulated chamber. This shell will be 4 inches (10 cm) thick. Double up panels of 2-inch (5 cm) rigid polystyrene foam. Use two layers for the lid, and if possible, use double layers on all four sides and the bottom. Glue panels together to achieve the ideal thickness.
5. Finish with a final radiant layer of foil or radiant barrier wrapped outside the completed box (which now consists of pot, insulation, chamber, insulation, polystyrene layer, final insulation layer).

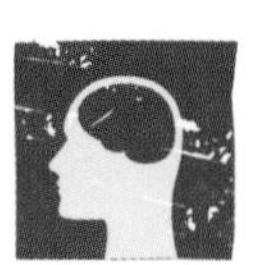

GOOD TO KNOW!

Makeshift cookers can be even simpler. You can pack crumpled newspapers, dry hay, blankets, comforters, down clothing, flannel, or closed-cell foam sleeping pads in a container around a heated pot of food. A wide-mouth thermos bottle, if it has a thick, well-insulated stopper, will also serve for this kind of cooking. Backpackers, for example, would preheat the unit with simmering water and then pour in boiling hot food before wrapping the container with insulators.

USING YOUR FIRELESS COOKER

To use a fireless cooker for most foods, except beans, bring the food to a rolling boil in a pot and then transfer it to the insulated cooking container. These dishes will be ready to serve after 2 hours or so of fireless cooking.

Beans are the exception to this rule because they don't soften readily if salty or acidic seasonings are added beforehand. Beans require a two-step process. First, soak the beans and rinse them. Then simmer them in a pot with water for at least 10 minutes with no salty or acidic seasonings prior to a 2- to 4-hour stint in the fireless cooker. Remove the pot, stir in the desired seasonings, and simmer for an additional 10 to 15 minutes, stirring occasionally, before returning the pot to the insulated cooker for another 2 to 4 hours.

One note about cooking grains: Because there's almost no boil-off using fireless cooking, you need to reduce the water (or other liquid) in your recipe by approximately one-eighth of the usual volume, though this is often a matter of experimentation.

REMEMBER FOOD SAFETY

Efficient cookers will maintain hot food for 8 to 12 hours without overcooking or deteriorating the contents. In my experience, cookers have maintained a 135°F (57°C) temperature after 12 hours. For the first two hours of slow cooking, a temperature in excess of 180°F (82°C) is normal. Don't allow foods to stand in the pot for extended periods below 140°F (60°C). Monitor the temperature with a meat thermometer, and try to keep food temperature at about 160°F (71°C) until served. Refrigerate leftovers promptly, especially meat and dairy dishes.

Return to Your Roots

Cost savings

About 19 cents per cup, comparing grocery-bought potatoes to boxed instant potatoes. Bulk potato purchases will lower the cost difference significantly

Benefits

Great-tasting and nutritious meals

Root vegetables are available just about everywhere, very economical, easy to keep and store, and quite nutritious. Except for potatoes, they're often neglected when planning meals, but these delicious recipes should inspire you to help root vegetables stage a big comeback.

Roots are good sources of calcium, potassium, and other trace minerals. Yellow root vegetables are high in vitamin A and beta carotene. And, best of all, they aren't just dull accompaniments these days, as the following recipes, provided and tested by *BackHome*'s food editor, Judy Janes, will prove.

From a "return on investment" standpoint, root crops are hard to beat, whether you grow them yourself or purchase them at market. Also, you can purchase root vegetables in bulk at a lower cost. These easy recipes offer a broad variety to suit diverse tastes.

Beet and Gorgonzola Salad

Makes 6 to 8 servings

- 8 to 10 beets
- 2 tablespoons (28 ml) white wine vinegar
- 3 tablespoons (45 ml) extra-virgin olive oil
- Salt
- Pepper
- ½ cup (60 g) walnut pieces
- ¼ pound (115 g) Gorgonzola

Wash the beets and trim off the stems and roots.

In a large kettle of boiling water, cook the beets until fork tender, about 30 minutes. Drain them and allow them to cool.

When the beets are comfortable to handle, peel off the skins with your fingers, and cut the beets into thin julienne strips. Place them in a medium-size bowl. Sprinkle with the vinegar and oil and salt and pepper to taste. Toss gently until coated. Cover and chill.

Preheat the oven to 350°F (180°C).

In a shallow pan, toast the walnuts for 5 minutes. Just before serving, carefully toss walnuts with the beets. Place in a shallow serving dish and crumble the Gorgonzola evenly over the top.

Parsnip Cakes with Roasted Red Pepper Sauce

Makes 2 to 4 servings

Parsnip Cakes

- 3 or 4 parsnips
- 1 egg, lightly beaten
- 1 cup (115 g) toasted, buttered bread crumbs
- 1 tablespoon (1.3 g) chopped fresh parsley
- Salt
- Pepper
- 2 tablespoons (28 g) butter or oil

Roasted Red Pepper Sauce

- 3 large red bell peppers (or 1 large jar roasted peppers, drained)
- 1 small onion, chopped
- 1 clove garlic, crushed and minced
- 2 tablespoons (28 ml) extra-virgin olive oil
- 1 large or 2 medium-size ripe tomatoes, chopped
- 2 tablespoons (5 g) chopped fresh basil or 2 teaspoons dried basil
- Salt
- Pepper

To make the Parsnip Cakes: Peel the parsnips and cut them into 1-inch (2.5 cm) pieces. Place them in a medium-size saucepan and cover with water to which a little salt has been added. Boil until tender, about 10 minutes. Drain the parsnips and return them to the pan over heat until all of the moisture evaporates. Allow the parsnips to cool, and then mash them and mix lightly with the egg, bread crumbs, parsley, and salt and pepper to taste. Chill for at least 30 minutes. (Meanwhile, make the sauce.) Form the mash into 4 cakes, and sauté them in a pan in the butter or oil until crisp, for 5 minutes each side. Dress with the sauce and serve immediately.

To make the Roasted Red Pepper Sauce: Preheat the oven to 500°F (250°C or gas mark 10).

Roast the fresh peppers directly on the oven racks for 20 minutes or so, turning to get all sides evenly charred, until the skins are black and blistered. Place the peppers in a covered bowl immediately. (Being covered and cooling in their own steam makes the peppers easier to peel.) When the peppers are cool, peel off the skins in large pieces, but do not wash. Remove the stems and innards and slice into strips. Set aside.

In a pan, sauté the onion and garlic in the oil for 5 minutes. Add the tomatoes, basil, and roasted pepper strips, and continue to cook to blend the flavors for 5 more minutes. Puree in a food processor or blender and season to taste with salt and pepper.

Carrot and Rutabaga Puree

Contributed by *BackHome* reader Bruce Howe

Makes 4 servings

- 4 or 5 medium-size carrots
- 1 rutabaga
- 2 tablespoons (28 g) butter
- ¼ teaspoon mace
- ½ cup (120 ml) milk, half-and-half, or sherry
- Salt
- Pepper

Scrape and peel the vegetables, and cut them into pieces.

In a medium-size saucepan, boil each lot separately until fork-tender. Drain, mash, and combine the cooked vegetables with the butter, mace, and milk, cream, or sherry. Add salt and pepper to taste.

Potatoes with Mustard and Chives

Makes 4 to 6 servings

- 6 to 8 medium-size new potatoes of uniform size (such as Red Bliss)
- 4 tablespoons (55 g) butter
- 2 tablespoons (30 g) Dijon mustard
- 2 teaspoons white wine vinegar
- 2 tablespoons (6 g) chopped chives
- Salt
- Pepper

Peel the potatoes into smooth round shapes. Place them in a medium-size saucepan and cover with cold water. Bring to a boil and cook for 10 minutes, or until barely tender. Drain the potatoes well, and return them to the pan over heat until all of the moisture evaporates.

Preheat the oven to 400°F (200°C or gas mark 6). Melt the butter in a heavy roasting pan or casserole. When the butter is bubbling but not brown, toss in the potatoes and coat evenly. Roast for 40 minutes, turning and basting occasionally, to ensure even browning and crisping on all sides.

In a small bowl, make a paste of the mustard, vinegar, chives, and salt and pepper to taste. Spoon it into shallow serving dish. Place the roasted potatoes on top, and lightly coat with the mustard sauce.

Try the Penny-Pincher's Guide to Dutch Oven Cooking

Cost savings

Preparing meals outdoors with charcoal briquettes or propane saves oven time.

Benefits

Cooking one-pot meals with low energy use or during power outages

Dutch ovens can be used to cook or bake just about anything, including prime rib, pizza, breads, and pies. They are versatile enough to use for roasting, stewing, baking, deep-frying, and boiling. Turn a lid upside down, and you can use it as a griddle or crepe pan to cook bacon and eggs, pancakes, or crepes. Here, contributor Gary F. Arnet outlines the uses of a Dutch oven and offers recipes, to boot.

The Dutch oven method is one of the simplest and most delicious forms of outdoor cooking. And since no electricity is required, this method comes in handy during electricity shortages or power outages.

Dutch oven cooking is easy to learn, but a few secrets will help you become an expert.

CHOOSING AN OVEN

With all the brands and styles available, buying a Dutch oven can be confusing. The first decision is: cast iron or aluminum?

Cast-iron ovens distribute heat better, retain it longer, and give a smoked flavor to foods. Cast iron reacts slowly to temperature changes, so foods don't burn as easily if the fire flares up. You can also overheat cast iron without warping or melting it. On the downside, cast iron requires more effort to maintain and will rust if left wet or exposed to water.

Aluminum ovens weigh about one-third as much as cast-iron units, they don't rust, and they are easy to maintain. They heat faster but distribute heat unevenly and cool rapidly. More coals are required to reach and maintain the correct temperature, and it is possible to melt aluminum if too much heat is used. Aluminum melts at around 1,100°F (593°C), whereas the melting point of cast iron is over 2,100°F (1,149°C). This isn't a problem if you're careful, but those who place the oven in a fire and forget it may be in for a surprise.

Size and features are the next consideration. Oven sizes vary from 8 to 22 inches (20.3 to 55.9 cm) in diameter, with 10- and 12-inch (25.4 and 30.5 cm) ovens being the most common. A 12-inch (30.5 cm) oven can be used to cook for 2 to 12 people and is often a good choice.

Buy an oven with three legs, each at least 1½ inches (3.8 cm) long. Legs prevent the oven from resting directly on the coals and help it produce more uniform heat. The lid should fit flush with the lip of the oven to avoid heat loss.

A loop handle should be in the center of the lid, and the wire handle on the oven itself (called the bail) should be easily movable and strong enough to lift an oven full of food. Invest in a quality Dutch oven that will last, such as Lodge, which has been making them since 1896.

MAINTAINING AN OVEN

When you buy a cast-iron oven, wash it with hot, soapy water to remove metal dust and wax coating placed during manufacturing. Dry it immediately, coat it inside and out with no-salt cooking oil, and place it in an unheated kitchen oven. Turn the oven on to 350°F (180°C or gas mark 4) for 2 hours, then turn it off and let the Dutch oven cool slowly in the oven. This "seasoning" will protect your Dutch oven. Aluminum ovens should be washed and dried but do not require seasoning.

After using a cast-iron unit, wipe out food with a paper towel and wash the oven with hot, soapy water. Dry it completely and wipe a light coat of cooking oil over the inside and outside to protect the seasoning.

Some people line their Dutch oven with aluminum foil or a disposable foil pan to avoid cleaning. Purists will insist the flavor is not the same, but because I hate to clean after cooking, I find this helpful at times. Foil is especially useful for recipes containing a lot of sugar, as it makes cleanup easier.

There are a few "don'ts" to remember with Dutch ovens: Don't scrape or scour the oven with metal utensils, wire scrubbers, or brushes, as this can remove the seasoning and allow food to stick to the oven. Although Dutch ovens are rugged, don't drop or handle them roughly, as they can break. Don't place cold water in a hot oven as it may crack or warp.

USING YOUR DUTCH OVEN

Controlling the temperature inside a Dutch oven is the secret that will make you successful from your very first meal. Any source of heat will work, from wood coals to propane stoves and even regular gas or electric ovens. When used outside, charcoal briquettes are the most reliable source of heat. Dutch oven cookbooks will list the number of briquettes you need for a temperature for each diameter of oven.

Most recipes require a temperature of 325°F (170°C or gas mark 3). A simple formula to obtain this is to take the diameter of the oven and add that many plus 3 briquettes to the top and that many minus 3 to the bottom. For example, when using a 12-inch (30.5 cm) Dutch oven, place 15 briquettes on top and 9 on the bottom for a reliable 325°F (170°C or gas mark 3) temperature. They should be spaced about 1 to 2 inches (2.5 to 5.1 cm) apart. About two-thirds of the heat will be on the top and one-third on the bottom. For cooking times of more than 1 hour, replace the briquettes about once an hour.

When baking bread or rolls, remove the Dutch oven from the bottom briquettes after about two-thirds of the cooking time has elapsed, leaving the top briquettes to finish the job. This will avoid burning the bottom of baked goods. Pies can be cooked by using a baking rack inside the oven to keep the pan about 1 inch (2.5 cm) above the bottom.

With the help of a good Dutch oven cookbook and a little practice, you will be impressing family and friends in no time. The following recipes are some of my family's favorites and are easy for first-time cooks.

Dutch Oven Pork Roast

If you prefer pork chops rather than a roast, you can use them in this recipe instead.

- 2 tablespoons (30 ml) olive oil
- 1 2-pound (0.9 kg) pork loin roast, scored
- 2 medium-size red apples, sliced
- 1 medium-size onion, sliced
- 5 garlic cloves, crushed
- 1 12-ounce (340 g) can sauerkraut, drained
- 1 cup (235 ml) white wine

Preheat a Dutch oven, and place the olive oil and pork roast in the oven. Cover the roast with the apples, followed by the onion slices, garlic, and sauerkraut. Pour the wine around the roast, and cover. Bake for 1 hour at about 350°F (180°C or gas mark 4). Add more hot briquettes to the top and bottom and bake for an additional 30 minutes, or longer if needed, until a meat thermometer placed in the thickest part registers 170°F (77°C). Makes 4 to 6 servings.

Simple Cherry Cobbler

Easy to make, this recipe is a tasty introduction to Dutch oven cooking and is especially good for teaching children. Any number of variations can be made on this recipe. Apple pie filling or canned peaches can be used instead of cherries. Chocolate instead of yellow cake mix can be used with cherry pie filling.

- 2 20-ounce (567 g) cans cherry pie filling
- 1 18-ounce (510 g) package yellow cake mix
- 1 12-ounce (355 ml) can lemon-lime soda

Line a Dutch oven with aluminum foil, and add the cherry pie filling. Sprinkle the dry, unmixed cake mix evenly over the cherry filling. Pour the can of soda over the cake mix, and gently stir, combining the cake mix and soda, but not the pie filling. Place the lid on the Dutch oven, and bake about 35 minutes at 325°F (170°C or gas mark 3) or until the top is brown and the cake springs back to the touch.

Stop Buying Bread

Cost savings

About 95 cents per loaf

Benefits

Delicious and healthful whole-grain breads

Baking bread from scratch is a luxury for most people in today's fast-paced society. But even the time-crunched and aspiring bakers can enjoy homemade loaves. *BackHome* contributor Heidi Gaschler shares how to turn out a savory 100 percent whole wheat bread that surpasses any store-bought white flour offering.

Ingredients are the key to baking successful loaves. Don't skimp. Choose whole wheat flour made from hard red winter wheat. This flour is high in protein, which converts into a gluten when mixed with a liquid. Elastic gluten stretches as bread rises. This action is the key to creating a "light" loaf.

Here are some other pointers to consider before you flour your hands.

- **Always purchase active dry yeast,** not a rapid-rise or quick-rise variety. Red Star or Fleischmann's are quality brands. Also remember that butter will produce lighter bread than oil—so go with butter.
- **Enhance yeast growth** and yield a high-rising dough by using water reserved from boiling potatoes in place of plain water. This trick is used by many seasoned bakers.
- **Don't try to eliminate salt from the recipe**. Besides enhancing the flavor, the salt keeps the yeast in check and prevents its over-rising and collapse of the bread.
- **Vigorously beat the "sponge"** (bread dough prior to adding all of flour) and allow it to rise. This step gives the yeast extra time to develop and allows the bran time to absorb liquid, decreasing the amount of flour required to form the dough into a ball. If you add all of the flour before the bran is saturated, you've added too much. The result is a dense bat. (And that means trouble for a "light" loaf.)
- **Use your muscles to knead the dough.** Go ahead and slam it around. Rough treatment of the bread dough develops the gluten, which needs to stretch during rising and baking. Well-developed gluten means bread is capable of rising well.
- **Don't rush the process**. We crave instant gratification and fast results. That's just not possible with bread baking. Most American bread recipes call for increased amounts of yeast and reduced rising times. This produces bread with a yeasty flavor that only increases after a day on the shelf. Leisurely rising times using less yeast give the wheat flavor time to develop.
- **Yes—you must take your time**. But you can still be a very busy person and bake bread. Just pick a day when you'll be home—doing chores, mowing the lawn, making a quick trip to the supermarket. These activities can take place while bread is rising.
- **If you grow impatient, punch the dough.** (This isn't a bad thing.) Specifically, punch down unrisen dough and allow it to rise again. This practice can be repeated several times without over-rising the tile bread, unless your house is unusually warm. Also, consider allowing bread to rise in the refrigerator overnight, as long as your refrigerator is 40°F (4°C) or less. Then, just shape it into loaves in the morning.

Whole Wheat Bread

Makes 4 loaves

- 5½ cups (1300 ml) potato water
- 1½ teaspoons active dry yeast
- 2 tablespoons (25 g) sugar or honey
- 2 tablespoons (30 g) salt
- ½ cup (112 g) butter, softened
- 10+ cups (1,200+ g) whole wheat flour

In a large pot or stockpot, heat the potato water to 115°F (46°C). Add the yeast and sugar or honey. Wait for the yeast to bubble. (The bubbling action will be slight given the high water-to-yeast ratio and may not be easily detectable with the naked eye. If you examine the mixture with a flashlight, you will be able to clearly perceive miniature bubbles rising to the surface.)

After the yeast bubbles, mix in the salt and butter. Add 6 cups (750 g) of the flour and beat for 2 minutes, or 200 strokes, with a large wooden spoon. Add 2 more cups (250 g) of flour, and beat an additional 200 strokes or 2 minutes. Cover and let rise for 1½ hours. Stir down and begin adding flour from the 2 cups (250 g) or more remaining.

Be careful not to add too much flour, as this will yield heavy bread. If the dough kneads well but sticks to your hands, grease them and the board with butter. Knead the dough vigorously for 10 minutes. Place the dough in a greased bowl and cover with a clean, damp towel.

Let the dough rise until it's doubled in bulk. (The rising time may vary from 1½ to 3 hours, depending on the temperature of your house.) Punch the dough down and form the dough into loaves, braids, butterhorns, etc. Place the dough in greased pans and cover it with a damp towel. Let the dough rise for 1½ to 3 hours, until it has doubled in bulk.

Bake the loaves at 375°F (190°C or gas mark 5) for 35 minutes for standard-size loaves. Bake miniature loaves and butterhorns for approximately 20 to 25 minutes. Remove the bread from the pans and cool on wire racks. Store the bread in the refrigerator, or freeze to preserve for a special occasion.

Make Winter Soups

Cost savings

A small savings per serving, but economy increases if they are served as whole meals.

Benefits

Healthful, hearty, homemade soups that are meals in themselves

Cold winter days call for a kettle of homemade soup simmering on the stove. Long, slow cooking brings out flavors, and the aroma stokes the appetite. *BackHome* contributor Patricia Rutherford knows a few things about northern winters, and the soups that go along with them. Read on for what she has to say.

Soup in every form warms the spirit. From traditional stews to a French pot-au-feu, the perfect soup is piping hot and can stand alone as a main course. Accoutrements might include crackers, pickles, and celery. Egg rivels also complement soup: Make them by working 1 unbeaten egg into 1 cup (125 g) of flour sifted, with 1/4 teaspoon of salt until you get a cornmeal-like mixture. Drop by the spoonful into boiling soup, cover lightly, and cook gently for 10 minutes. Rivels separate and look like rice.

Italian Vegetable Soup

Makes 8 to 10 servings

- 1 pound (455 g) ground beef
- 1 large onion, chopped
- 1 cup (100 g) chopped celery
- 1 cup (130 g) chopped carrots
- 2 cloves garlic, minced
- 1 15-ounce (425 g) can tomatoes, with juice
- 1 15-ounce (425 g) can tomato sauce
- 1 16-ounce (454 g) can small red beans, with liquid
- 1 cup (235 ml) water
- 5 teaspoons beef bouillon granules
- 1 tablespoon (4 g) fresh parsley
- 1 teaspoon salt
- ½ teaspoon dried oregano
- ½ teaspoon dried basil
- 2 cups (140 g) shredded cabbage
- 1 cup (100 g) green beans
- ¼ to ½ cup (26.3 to 52.5 g) macaroni

In a 5-quart (5 L) pot, brown the ground beef; add the onion, celery, carrots, garlic, tomatoes, tomato sauce, beans, water, bouillon, parsley, salt, oregano, and basil. Cover and simmer for 30 minutes. Add the cabbage, green beans, and macaroni. Cook until done. If the soup is too thick, add water.

Cheese Soup

Makes 5 to 6 servings

- 1 medium-size potato, peeled and diced
- 1 medium-size onion, diced
- 2 chicken bouillon cubes
- 1 quart (946 ml) water
- 1 10-ounce (280 g) package broccoli, cooked
- 1 10 ¾-ounce (303 g) can cream of chicken soup
- ½ pound (225 g) Velveeta

In a stockpot or Dutch oven, cook the potato, onion, bouillon, and water for about 30 minutes, then add the broccoli, soup, and cheese. Continue over medium heat until the cheese is melted.

Creamy Mushroom-Broccoli Soup

Makes 6 servings

- 4 tablespoons (55 g) butter
- 1 medium-size onion, chopped
- 12 ounces (340 g) fresh mushrooms, sliced
- ¼ cup (30 g) flour
- ½ teaspoon salt
- ¼ teaspoon white pepper
- 1½ cups (350 ml) milk
- 1 10-ounce (285 ml) can chicken broth (or vegetable broth)
- 1 cup (235 ml) half-and-half
- 1 10-ounce (280 g) package frozen chopped broccoli, cooked and drained

In a stockpot or Dutch oven, melt the butter over medium-high heat. Sauté the onion for 2 or 3 minutes, then add the mushrooms and cook for 2 to 3 minutes. Sprinkle on the flour, salt, and pepper, and stir. Lower the heat, add the milk and broth, and cook, stirring, until thickened. Continue heating but do not let boil, while adding the half-and-half and broccoli. Heat to serving temperature, then serve immediately.

Chicken Noodle Soup

Makes 8 to 10 servings

- 1 2- to 3-pound (910 g to 1.4 kg) stewing chicken
- 2½ quarts (2.3 L) water
- 3 teaspoons salt
- 2 chicken bouillon cubes
- ½ medium-size onion, chopped
- ⅛ teaspoon ground black pepper
- ¼ teaspoon dried marjoram
- ¼ teaspoon dried thyme
- 1 bay leaf
- 1 cup (130 g) diced carrots
- 1 cup (100 g) diced celery
- ½ cup (70 g) uncooked fine noodles

Place all the ingredients except noodles in a large soup pot. Cover and bring to a boil; skim the fat from the broth. Reduce the heat, cover, and the simmer for 1½ hours, or until the chicken is tender. Remove the chicken from the broth and cool. Debone the chicken and cut it into chunks. Skim the fat from the broth, and bring it to a boil. Add the noodles; cook until done. Return the chicken to the pot; adjust the seasonings to taste. Remove the bay leaf.

Cream of Peanut Soup

Makes 6 servings

- 1 medium-size onion, chopped
- 2 ribs celery, chopped
- ¼ cup (55 g) butter (½ stick)
- 3 tablespoons (24 g) flour
- 2 quarts (1.9 L) chicken or vegetable stock
- 2 cups (520 g) smooth peanut butter
- 1¾ cups (420 ml) light cream
- Chopped peanuts

In a soup pot, sauté the onion and celery in butter until soft but not brown. Stir in the flour until well blended. Add the stock, stirring constantly, and bring to a boil. If a smoother soup is desired, remove from the heat and rub through a sieve. Add the peanut butter and cream, stirring to blend thoroughly. Return to low heat, but do not boil. Serve garnished with peanuts.

Soupmaking Suggestions

Use a large pot. A too-small pot can bring on spillovers or cause the mixture to heat too slowly.

Don't cheat on cooking time. Simmering soup over low heat helps to extract the maximum flavor from the ingredients.

Take advantage of technology. Food processors, blenders, and salad makers are useful in preparing soups. Use them to chop or puree vegetables.

Flavor up with bouillon. Add a bouillon cube to cream soup or weak homemade stock to enhance flavor.

Skim the fat. Remove fat from stock or soup by chilling it and then removing the solid layer, or by rolling a paper towel to skim small amounts of fat from the surface of warm soup. For larger amounts, use a bulbed meat baster.

Heat dairy-based soups slowly. To prevent separating, do not allow soup to come to a boil.

Save the leftovers. Most soups can be covered and refrigerated up to 3 days. However, those made with fish or shellfish should be stored for 1 day at the most.

Stretch your budget. Save leftover vegetables and vegetable juices for soupmaking.

Mind the seasoning. Some soups when frozen may undergo flavor or textural changes. Certain vegetables, such as bell peppers, intensify in flavor if frozen, and onion gradually loses strength. After reheating soup, adjust the seasoning to taste.

Just add liquid. Dense soup tends to thicken during storage. When reheating, add a little broth, milk, or half-and-half until the desired consistency is reached.

Fill Up on Stews for Less

Cost savings

About $54 when compared to a similar meal dining out

Benefits

The satisfaction of preparing a delicious, nutritious, inexpensive, soul-satisfying meal yourself at home

Stew is a soul-warming supper for winter days. These stews are slow-cooked, consist of simple, down-home ingredients, and are sure to fulfill your appetite. *BackHome*'s food editor Judy Janes offers these stews for you to sample.

Frogmore Stew

The old Gullah culture is being preserved on St. Helena, one of the islands off the South Carolina coast. Frogmore, not much more than a crossroads, is the island's center of activity, where Frogmore stew is proudly offered in several variations. Our version is derived from a recipe given us by the Gay Fish Company, where the shrimp boats tie up to deliver the day's catch.

Makes 6 servings

- 1 pound (455 g) mild sausage, cut up
- 2½ tablespoons (35.4 g) butter
- 1½ tablespoons (9 g) Old Bay seasoning
- 1½ teaspoons salt
- 1 large onion, diced
- 3 ears corn, halved
- 6 new potatoes, peeled and diced
- 1 pound (455 g) medium shrimp (40 to 50 to a pound)

In a pan over medium heat, brown the sausage. Half-fill a 6-quart (5.7 L) stockpot with water. Add the butter, seasoning, salt, sausage, and onion. Bring to a boil. Add the corn and potatoes and boil for 4 minutes. Add the shrimp and simmer for 4 to 6 minutes, or until the shrimp float. Remove from the heat, and let sit for 4 minutes. Drain off all but 2 cups (475 ml) of liquid.

Serve immediately, accompanied by butter and/or cocktail sauce and lots of napkins.

Chicken Marengo

Legend has it this dish was created to celebrate Napoleon's narrow victory over the Austrians at the Battle of Marengo on June 14, 1800. Napoleon's chef, claiming he didn't have much to work with, put on his genius cap and created this classic.

Makes 6 servings

- 2 cups (475 ml) Sauce Espagnole (see page 45)
- 1 tablespoon (15 ml) olive oil
- 2 tablespoons (30 g) butter
- 1 roasting chicken, cut into pieces
- 1 large can (28 ounces [794 g]) tomatoes, diced and undrained
- 1 tablespoon (16 g) tomato paste
- 1 teaspoon garlic salt
- ½ cup (120 ml) white wine
- ½ pound (225 g) mushrooms, cleaned and trimmed (save stems for sauce)
- Parsley
- Toast triangles, with the crusts removed

Make the Sauce Espagnole and keep warm.

In a stockpot or flameproof Dutch oven, heat the oil and butter, and thoroughly brown the chicken. Remove the chicken with a slotted spoon and drain on paper towel. Add the tomato, tomato paste, and garlic salt, and cook, stirring constantly, for 2 to 3 minutes. Add the wine, and cook at a slow boil until the sauce is reduced by half. Add the Sauce Espagnole and mix well. Add the chicken and mushrooms, cover, and simmer for 30 minutes, or until the chicken is tender. Just before serving, move to serving platter. Sprinkle with parsley and garnish with toast triangles.

Sauce Espagnole

Makes 2 cups (475 ml)

- 6 tablespoons (85 g) butter
- 1 small onion, peeled and finely diced
- 1 small carrot, peeled and finely diced
- 1 celery stalk, diced fine
- 3 tablespoons (48 g) flour
- 1 teaspoon tomato paste
- Chopped stems of mushrooms (reserved from Chicken Marengo)
- 4 cups (950 ml) beef broth (or homemade stock)
- 1 teaspoon fines herbes mix
- Salt
- Ground black pepper

In a large nonreactive saucepan, heat the butter and add the onion, carrot, and celery. Cook until the vegetables just begin to brown. Add the flour and stir with wire whisk to prevent lumping and burning. When the mixture is a rich dark brown, remove it from the heat and allow it to cool slightly. Stir in the tomato paste, the chopped mushrooms, 3 cups (705 ml) of the beef broth, and the herbs.

Bring the mixture to a boil, stirring constantly. Reduce the heat, and cook for 35 minutes, stirring occasionally, to prevent sticking. Add beef broth as necessary to keep the sauce smooth and silky. Salt and pepper to taste. Strain the mixture, pressing the vegetables to extract their juices. (The sauce should have the consistency of heavy cream.) Return the sauce to the pan and keep warm for use.

This sauce keeps covered in the refrigerator for 1 week and in the freezer for several months.

Brunswick Stew

Named for Brunswick County in southeastern Virginia, this originated as a hunter's stew made of squirrel or rabbit and onions. Nowadays, it usually consists of chicken and vegetables and sometimes game.

Makes 6 servings

- 1 roasting chicken, cut up
- 4 cups (950 ml) water
- 1 teaspoon salt
- 1 teaspoon dried dill
- 4 slices lean bacon
- 1 large onion, chopped fine
- ½ pound (225 g) ground chuck, cut into cubes
- 1 teaspoon paprika
- 28 ounces (794 g) diced and undrained tomatoes
- 1 box (10 ounces [280 g]) frozen lima beans or 1½ cups (255 g) fresh lima beans
- 1 box (10 ounces [280 g]) frozen whole-kernel corn or 1½ cups (150 g) fresh corn, cooked
- 3 or 4 potatoes, peeled and diced
- Salt
- Ground black pepper
- Flour

In large stockpot or flameproof Dutch oven, cover the chicken with the water and add salt and dill. Bring to boil and cook, covered, for 1 hour, or until tender. Cool. Remove the chicken from the bones, reserving the broth, and cut the chicken into bite-size pieces. Dice the bacon, and cook it thoroughly in the same pot. Remove the bacon, add the onion, and cook until soft. Push the onion aside; add the beef and brown well on all sides.

Add the bacon, reserved chicken broth, paprika, and tomatoes. Bring to a boil, cover, and simmer for 1½ hours. Add the lima beans, corn, potatoes, and chicken, and simmer for 30 minutes. Season to taste with salt and pepper, then thicken slightly with a flour-and-water paste. (Blend ¼ cup [31 g] of flour with ¼ cup [60 ml] of water with a fork until there are no lumps.)

Vegetarian Stew

This stew features lots of vegetables and is topped with a thick broiler-browned crust of mashed potatoes. After the first three ingredients, the rest is up to the cook.

Makes 6 servings

- 4 medium-size onions, sliced
- 3 tablespoons (45 g) butter
- 2½ cups (590 ml) vegetable stock
- 4 or 5 potatoes, peeled and cut into chunks
- 2 carrots, peeled and cut into 1-inch rounds
- 2 cups (240 g) chunks of winter squash
- 1 cup (150 g) peas
- ½ cup (50 g) green beans, cut into 2-inch (5 cm) pieces
- 1 cup (150) corn, fresh or frozen
- 2 cups (140 g) sliced fresh mushrooms
- 1 large can (28 ounces [938 g]) crushed tomatoes
- 1 teaspoon chopped fresh basil
- 1 cup (235 ml) red wine
- Salt
- Ground black pepper

In a large stockpot or Dutch oven, sauté the onion in butter. Add the stock. Reduce the heat and simmer for 50 minutes. Add the potatoes, carrots, and squash, and cook for 15 minutes. Add the peas, green beans, corn, mushrooms, tomatoes, basil, and red wine. Season to taste with salt and pepper. Continue cooking for 15 minutes.

Blend Your Own Spices

Cost savings

Typically 30 or 40 percent of the cost of comparable store items

Benefits

Fresher, better-tasting blends at a bargain price and save time when preparing recipes you make often

There are a bounty of spice blends on market shelves today, ranging from Super Salad Spices to Ragin' Cajun Seasoning to Salt-Free this and that. Blends are convenient, but you'll pay extra for them. Plus, they might contain unexpected amounts of salt (no!), sugar, or monosodium glutamate (MSG) in one form or another. The answer to avoiding these problems is simple: Make your own. Here, contributor Sandra K. Bowens offers recommendations for guaranteeing fresher taste, known ingredients, and a reasonable cost.

I started making my own spice blends for camping trips to avoid packing all of the individual bottles. Now I have more than ten different seasoning blends in my spice cabinet, each one made to suit my own tastes.

The best part is that they have salt only if I want them to. If you read the labels on many commercial seasoning combinations, you'll see that the first ingredient is salt. Salt is one of the heaviest, not to mention most inexpensive, flavorings, so manufacturers often use this for filler. Flavored salts on the supermarket shelves, such as garlic salt and onion salt, are sometimes as much as 90 percent salt to 10 percent flavoring, but you pay a premium for the flavoring.

MSG is a heavy filler and, if you are like me, an unwanted additive. Watch for hydrolyzed proteins (other "flavor enhancers") that are often used in place of MSG. I came across a vegetable blend with sugar in it; what could that be but a filler? Common sense tells me that flavor enhancers are not necessary if you start with quality ingredients.

You will get a fresher-tasting blend if you make your own from dried herbs and spices you buy in bulk. Look for purveyors who specialize in them, because their supplies will turn over more quickly. Grind your own whole spices just before blending for more freshness. You can do this with an electric spice grinder or a mortar and pestle. Keep blends fresh in containers with tight-fitting lids, and store them in a dark place as far away from heat sources as possible.

For most blends I prefer to make a small quantity right in the jar I plan to store it in. I simply measure the specified amounts into the container, then shake well. It's that easy. You could also mix the ingredients in a glass or ceramic bowl, funneling the blend into the jar. Avoid plastic mixing bowls because they will sometimes stain or create static electricity. For these same reasons, if your funnel is plastic, you might prefer to use a rolled-up sheet of paper instead.

It is a snap to create blends from your favorite recipes. For the sake of convenience, I typically multiply the seasonings by four or six so that I have a blend for future use. When preparing the recipe, total up the measurements from all the herbs and spices and then add that amount of the spice blend. For example, if the original recipe calls for 1 teaspoon tarragon, 1/2 teaspoon dry mustard, and 1/4 teaspoon pepper, multiply these amounts by four. Measure 4 teaspoons tarragon, 2 teaspoons dry mustard, and 1 teaspoon pepper into a jar and shake well. Next time you make the recipe you would use 1¾ teaspoons of the blend.

The following recipes have worked for me for many years. You can use them as written or to spark ideas, or tinker with them to meet your own taste requirements. I came up with the chili powder blend by simply premixing all the herbs and spices that I had been adding individually to my own chili. Now, rather than opening and closing six bottles every time, I open one and use as much as needed.

Whether you use these recipes or create your own blends, you will enjoy knowing that the ingredients are fresh, the additives are gone, and the cost is about half. One Web site offered two-ounce blends for $4.50. I calculated the cost for two ounces of my Italian herbs to be $2.86. Now that's what I call incentive.

Basic Beef Blend

This goes well anywhere beef does, such as in stroganoff or soup, on steaks, or as rub for ribs.

- 3 teaspoons leaf marjoram (dried)
- 2 ½ teaspoons whole thyme (dried)
- 2 teaspoons granulated garlic
- 1 teaspoon ground black pepper
- 1 teaspoon salt
- ½ teaspoon dry mustard

Measure all ingredients into a jar and shake well.

Basic Chicken Blend

I think of this as my own poultry seasoning.

- 3 teaspoons sage, ground or cut and sifted
- 3 teaspoons whole thyme (dried)
- 2 teaspoons hot Hungarian paprika
- 1 teaspoon salt

Measure all ingredients into a jar and shake well.

Basic Pork Blend

Use this to season grilled pork chops, mix into ground pork for a quick sausage, or combine with olive oil for a marinade.

- 1 teaspoon ground sage
- 1 teaspoon whole thyme (dried)
- 1 teaspoon ground black pepper
- 1 teaspoon granulated garlic

Measure all ingredients into a jar and shake well.

Potato Spice

This is excellent with pan-fried potatoes or with boiled potatoes from which most of the cooking water has been drained. Add plenty of butter and plenty of the blend.

- 2 tablespoons (2.6 g) dried parsley
- 1 teaspoon paprika (hot or sweet)
- ¾ teaspoon ground black pepper
- ¾ teaspoon garlic salt
- ½ teaspoon onion salt
- ½ teaspoon dill weed (dried)

Measure all ingredients into a jar and shake well.

Economize Using Hand-Ground Grains

Cost savings

More than $1.50 per bread loaf; between $4 and 6.50 per week on cereals, depending on family size

Benefits

Healthy staples at an economy price if bought in bulk

For years, health-minded people have praised the merits of a diet rich in whole foods, and today even mainstream nutritionists are advising us to replace white bread with a variety of whole grains and whole grain products. Home grinders can help incorporate these important foods into our diets.

Whole grains, such as brown rice, wheat berries, millet, and rye, can all be boiled and eaten like white rice, but many people complain about the lengthy cooking times and boring results. With a grinder, these foods can be cracked to greatly reduce preparation time. For example, whole wheat berries require an hour or more of cooking, whereas cracked wheat can be ready in 10 minutes. Cracking also opens up myriad possibilities, including puddings, soups, cereals, salads, and pilafs, and flour in baked goods.

Furthermore, the bulk purchase and safe storage of staple and even exotic grains is remarkably inexpensive when compared to buying the breads, cereals, and so forth ready to eat from the store. Here, *BackHome* contributor Susan Grelock looks at the benefits of hand-grinding for home use.

With a little tenacity and imagination, I've incorporated my grain mill into everyday life, and it's now an indispensable part of my kitchen. If you've ever tried grinding a hopper of grain by hand, no doubt you worked up quite a sweat while asking the obvious question, Why bother?

The health reasons are obvious, but a home mill also allows you to crack grain only as you need it. Even though cracked grains and flours are available in stores, they're best cracked at home shortly before being used, to prevent spoilage. A whole dried corn kernel, for instance, will keep for years in its natural state, but once it's cracked the oils quickly begin to oxidize and turn rancid. A home grinder gives you access to the more nutritious and sweeter-tasting fresh flours.

Cracking your own means you'll be purchasing whole grains. These are almost always less expensive than milled grains. Because they store well, you can buy them in large quantities at an even more appreciable savings. I purchase a 50-pound sack of organic wheat berries from my local co-op for 27 cents a pound and store them in plastic buckets.

You can buy old mills at flea markets and yard sales; newer models—and replacement burrs and parts—are sold through health food stores and mail-order sources. Up until a few years ago, there wasn't any question about what kind of grinder to get—they were all stone grinders. Today, the introduction of steel-burred grinders has opened up a few options.

THE DAILY GRIND

All of these perks don't change the fact that if you want to grind grains by hand, you're going to have to work at it. One of the keys is timing. Three loaves of wheat bread require 8 cups (1 kg) of flour; that's about two full hoppers (which you'll most likely want to run through twice to get a fine bread flour). This is not something to save for the last minute. Start grinding the day before you want to bake, spacing out the task during the day. I like to fill the hopper while I'm in the house working. I'll grind just a little, perhaps 20 turns, bring in firewood, grind a little more, sweep the house. Before I know it, I have a big bowl of flour.

In our early days with our Corona grinder, I used it primarily for cracking small amounts of grains for hot cereal and other dishes. I advise anyone getting familiar with their mill to start by making cereal. Any whole grain can be cracked and boiled in water to produce an excellent porridge. (See Basic Hot Cereal recipe on page 51.) Even if you are not a hot "mush" fan, you might find freshly cracked cereals make a sweet, nutty porridge more enjoyable than packaged varieties.

You can also steam very coarsely cracked cereal to produce a quick-cooking dish to use like rice. This can be chilled and tossed with vegetables and vinaigrette to make a refreshing pilaf.

As you begin to substitute your own flour in recipes, remember that you're making whole wheat flour, which will produce a different result than white flour. I have always substituted whole wheat for white flour in recipes and find that it give baked goods a deeper color and more interesting flavor, as well as extra nutrition. In working with hand-ground flour, you probably will never get it as fine as commercial flour. This means that it will be a little slower to absorb liquid in a dough. If a batter seems a little wetter than usual, let it sit for 5 minutes before adding more flour. For bread dough, you'll probably have to knead a little longer than usual to help the flour absorb the liquid.

Of course, wheat is not the only grain worth cracking. A trip down the bulk aisle in a natural foods store will reveal a wide range of grains.

Wheat berries: Hard, soft, winter, spring, white, and red berries all make a good all-purpose flour. Soft wheat is best for pastry flour, suitable primarily for cookies and cakes.

Spelt and kamut: These are ancient varieties of wheat that have been rediscovered. They are subtly different from regular wheat but can be used interchangeably.

Rye berries: Whole rye flour turns rancid very quickly, and for this reason it is not generally produced commercially ("dark" rye flour is not whole). Grind your own as you need it, and enjoy the distinct, rich taste. Rye berries are soft and therefore difficult to grind, but because the strong flavor goes a long way you'll need to add only a little to whole wheat batters. To make it easier to grind, mix rye in with wheat berries and mill them together.

Corn: Whole corn for grinding is available in blue or yellow, both used the same. Don't use popcorn, which has been bred to have a thick hull. I could never go back to commercial cornmeal once I began making my own corn flour. Its sweet, fresh taste is vastly superior.

Brown Rice: There are generally three types of brown rice available: sweet, short, and long. Sweet rice is distinctly sticky and dense, so I prefer short and long grain for cooking. Brown rice makes a wonderful breakfast porridge that children, in particular, enjoy, and it's especially good in pudding.

Millet: Millet is a staple grain in Asia and Africa, and because of its nutritional superiority is sometimes called the queen of grains. The tiny golden balls grind easily (a bonus). A low gluten content makes millet flour suitable only in small amounts in yeasted dough, but you can substitute a little for the flour in cookie, muffin, and pancake recipes. Millet also makes a wonderful breakfast porridge that is fluffy and light; when made thick it can be used as a side dish similar to mashed potatoes or as a bed for saucy bean dishes.

Stock up on a variety of grains and dust off that grinder. There's a whole new world of grain cooking available if you're willing to simply "lend a hand."

Basic Hot Cereal

Follow these general instructions for a thick porridge that makes a great breakfast cereal and dinner side dish. Experiment with a variety of grains and blends; you'll discover the diversity of flavors and textures among grains. Crack the chosen grain as desired; a fine grind will produce a creamy, smooth cereal, whereas a coarser grind will yield a chewier cereal.

Makes 4 servings

- 2 cups (475 ml) water
- 1 cup (100 g) cracked grain
- Salt

In a medium-size saucepan, stir together the water, grain, and salt. Bring to a boil, stirring occasionally with a whisk to prevent lumps. Reduce the heat to low, and cover. Stir occasionally and cook until the grain is tender, about 10 to 20 minutes, depending on the coarseness of cracked grain.

Variation: If you prefer a less creamy cereal, try gently sautéing the cracked grain in a little oil before adding the water.

Grow Your Own Sprouts and Save

Cost savings

Bought in 5-pound (2 kg) bulk, organic alfalfa sprout seeds cost about 45 cents an ounce; enough to make around $16 worth of grocery-bought sprouts

Benefits

Freshness, low cost, and a wide selection of sprout choices

Sprouts are tasty, versatile, and packed with vitamins and minerals—A through E plus G and K, calcium, magnesium, potassium, and others. Even better, sprouts grow fast and cost about 10 to 30 cents a pound. You can grow sprouts quickly with the bottle-gardening method. Here's how to do it.

MATERIALS

- Seeds (1 tablespoon [7 g] small, such as alfalfa; or 3 tablespoons [21 g] large, such as mung beans)
- Wide-mouthed quart-size (1 L) canning jar
- Cheesecloth or nylon mesh
- Rubber or metal ring
- Wide bowl or glass baking pan

STEPS

1. **Prepare the seeds.** Put about 1 tablespoon (7 g) of small seeds or 3 tablespoons (21 g) of large seeds in a wide-mouthed quart-size (1 L) canning jar. Rinse the seeds thoroughly in fresh water, picking off any hulls or chaff. Pour off the rinse water, add about three times as much tepid (70° to 75°F, or 21° to 24°C) water as you have seeds, and put the jar in a warm, dark place overnight to allow the seeds to soak.

2. **Flush the seeds.** Fasten cheesecloth or nylon mesh over the jar's open mouth with a rubber band or metal ring. Pour out the water (you can save it for soup stock). Flush the seeds with fresh water, then pour out the water.

3. **Set the jar aside.** Prop the jar, mouth down, inside a wide bowl or glass baking pan, so that any excess liquid will drain away from the seeds. Place the bowl and jar in a warm, dark place.

4. **Repeat fresh-water rinse.** Repeat steps 2 and 3 two or three times each day to keep the seeds fresh. Sprouts will "sprout" in three to six days. Alfalfa sprouts are generally ready when they are about 2 to 3 inches (5 to 7.5 cm) long, and mung bean sprouts should be 3 to 4 inches (7.5 to 10 cm) long. Grain sprouts should be only barely longer than the seed itself.

5. **Green your sprouts.** You can set near-ready sprouts in a semibright window for a few hours to "green" them. But be careful not to leave them in the sun too long or they will lose their nutritional power. (So much for typically bright green store-bought sprouts!) Better to leave the sprouts whitish or only barely green.

Make Herbal Vinegars

Cost savings

Between $2.50 and $9.00 per 12-ounce (355 ml) bottle, depending on how expensive your tastes are

Benefits

Flavorful and healthy vinegars for a fraction of the cost of store-bought

If your experience with vinegar has been limited to the plain cider variety, you're missing a culinary treat. Herbal vinegars add an exciting taste to any food requiring vinegar, whether it's vinaigrette dressing, meat marinade, fresh salad, or the old-fashioned pioneer favorite, vinegar pie. Herbal vinegars are also great for sharing with friends and, when presented in attractive bottles, they make pretty and practical gifts. Here, *BackHome* contributor Marcella Shaffer walks us through the basics of creating herbal vinegars.

All herbal vinegars start with a base vinegar. It's best to use the more subtle and delicate varieties so they will not overpower the taste of the added herbs. Vinegars made from red and white wine, rice wine, and champagne are all good choices. Plain white distilled vinegar is good to use for more robust applications. When shopping for vinegars, it is not necessary to buy expensive brands, but white or colorless vinegars show off the ingredients inside the bottle, adding to the overall aesthetic value.

Many enthusiasts choose to make their own base vinegars from scratch, using fruits, wine, barley, and so forth, and saving the "mother" from each batch for the next. These homemade vinegars will work fine as a base for your herbal vinegars

UNDERSTANDING THE BASICS

There are two basic methods used in making herbal vinegars. The first is to place the herb in the bottle, fill it with base vinegar that has been boiled and cooled, and leave it to steep at room temperature for two or three weeks. After two weeks, taste for piquancy by putting a few drops on a piece of bread. After three weeks, if the flavor is not distinct enough for you, add more ingredients.

The second method produces faster results. Begin by bruising or crushing the flavoring ingredient, using use a pepper mill, coffee bean grinder, dull knife, or garlic press. Place it in a canning jar or other such container. Heat the base vinegar to boiling, and then pour it into the jar and cover it. Let the vinegar steep at room temperature. Some vinegars take just a few hours, while others may take a few days. Sample your vinegar for taste. When it has reached the desired strength of flavor, strain it though a coffee filter to remove debris, and pour it into the bottle. Add a small amount of the fresh ingredient for looks.

All vinegars will keep indefinitely at room temperature, though you should avoid storing your vinegar in direct sunlight. Although not absolutely necessary, boiling the base vinegar before using, as in the first method, will keep the "mother" from foaming and turning cloudy.

Note: Because vinegar is high in acid, it does not generally support the growth of bacteria. However, some may still allow bacterial growth, especially if the containers or herbs are not cleansed properly. For more information on preparation and storage, visit cecalaveras.ucdavis.edu/vinegar.htm.

FINDING BOTTLES

When making herbal vinegars, go for looks as well as flavor. You can buy bottles designed especially for herbal vinegars from many sources selling herbs and other culinary supplies. Search for attractive or unique bottles at flea markets, antiques shops, and yard sales. Any bottle that can be washed and can be tightly sealed with a lid or cork will work. If the lid is metal, place a small piece of plastic wrap between the lid and the vinegar to prevent a reaction from the acid in the vinegar. You can buy inexpensive corks of several sizes at stores selling beer-making supplies.

Think *recycled* when searching for bottles. Look around your kitchen for empty sauce bottles such as those from Tabasco or Worcestershire, large-size containers from vanilla or other flavorings, and beer and wine bottles.

Soak the containers to remove the labels, and then replace these later with your own labels. Before using the containers, wash them thoroughly and sterilize them by filling them with boiling water. Let this sit for a few minutes, then pour it out and invert the bottles on a towel to dry. When wrapped in tissue paper and tied with raffia or ribbon, a "found" bottle filled with your own herbal vinegar makes a very presentable gift.

MAKING HERBAL VINEGARS

There are no hard-and-fast rules in making herbal vinegars; be guided by your own tastes and preferences. Use herbs whose flavors appeal to you and that are available fresh, either from your own garden or from the market. Don't be afraid to experiment. You can make herbal vinegar with just a single herb or a combination of two or more. You can add allspice, dill, celery seed, peppercorns, or other spices for extra flavor; cinnamon sticks and spirals of lemon peel add zest as well as an attractive appearance.

To make a simple herbal vinegar, select the herb and rinse it under running water to remove dirt, and then pat it dry. Choose a bottle and prepare it, as described at left. For dill, tarragon, basil, thyme, and chervil, generally plan to use two large sprigs for each cup of base vinegar. Chive vinegar, however, will require more than just two stalks. If the flavor is not as strong as you like, add more of the herb (or herbs) and let it steep some more. If the flavor is too intense, dilute the mixture with more base vinegar.

USING HERBAL VINEGARS

Try adding basil vinegar to your favorite baked chicken or beef stew recipe, or sprinkle some over fresh tomatoes. Oregano vinegar is delicious on tomatoes, in pasta salads, or in almost any Italian dish. Cilantro or coriander vinegar adds rich flavor to salsa and other Mexican dishes. Any herbal vinegar is very good in homemade mayonnaise and potato salad.

Save Energy with Combination Cooking

Up to $1 per meal

Benefits

Saving time and energy cost

Leverage the energy-saving potential of your oven by cooking an entire meal at one time—the ultimate in dinner multitasking. This takes planning and preparation, but after you've done it once or twice, you'll look forward to that last relaxed hour or so before dinnertime.

Start by choosing a menu, then assemble the baking dishes and arrange them in an unheated oven. Position them according to how much attention you must give the dish during cooking. (Set a dish that needs an occasional look or special attention, or one that takes less time than the others, toward the front.) Next, save time by gathering all ingredients before you begin preparing recipes.

The menus we've chosen include main dishes, side dishes, and desserts. Mix and match as you please, as every recipe is baked in a preheated 350°F (180°C or gas mark 4) oven.

Marbled Meat Loaf

Makes 6 to 8 servings.

- 1 large onion, peeled and thinly sliced
- 2 tablespoons (28 g) butter
- 1 teaspoon (6 g) garlic salt
- 2 cups (100 g) fresh bread crumbs (see sidebar, page 58)
- 1 10-ounce box (300 g) frozen chopped spinach
- 2 tablespoons (30 ml) soy sauce
- Freshly ground black pepper
- 2 pounds (900 g) lean ground beef

Cook the onion in the butter over medium heat, stirring frequently so it doesn't burn, until a dark brown color. Chop, then mix with the other ingredients until well blended. Shape into a rounded oval and place in an oiled oven-proof dish. Bake for 1 hour at 350°F (180°C or gas mark 4). Serve with Rich Brown Sauce (see page 58) if desired.

Carrot Loaf

Makes 6 or more servings.

- 2 cups (450 g) carrots, peeled and chopped
- 1 cup (225 g) cottage cheese
- 2 eggs, beaten
- 1 tablespoon (15 g) flour
- 1 tablespoon (14 g) butter
- 2 tablespoons (30 g) packed brown sugar
- 2 teaspoons (2.6 g) chopped fresh parsley
- 1 teaspoon (6 g) salt
- Pinch of nutmeg

Cook the carrots, covered with water, until tender, about 15 minutes. Place in a blender or food processor with the remaining ingredients, and process until smooth. Pour into a buttered loaf pan, and place into a larger pan with water halfway up the sides of loaf pan. Bake for 55 to 60 minutes at 350°F (180°C or gas mark 4), until a knife inserted in the middle comes up clean. Remove to a rack to cool slightly; invert onto a platter.

Chicken Tetrazzini

Makes 6 servings.

- 1 cup (225 g) fettuccine pasta, cooked and drained
- 1 recipe Velouté sauce (see page 58)
- 4 tablespoons (55 g) butter
- 3 skinned and boned chicken breasts, halved
- 2 cups (140 g) sliced mushrooms
- ½ teaspoon (1.1 g) ground nutmeg
- ½ cup (±120 ml) white wine
- ½ cup (50 g) freshly grated Parmesan

Butter a shallow 2-quart (2 L) baking dish. Toss the cooked and drained fettuccine with the sauce, and spread in dish. Melt 2 tablespoons (28 g) of the butter in a skillet and lightly sauté the chicken, browning on all sides. Arrange on top of the fettuccine.

Melt the remaining butter in a skillet, add the mushrooms, and toss to coat. Cook on high heat until the mushrooms brown around the edges. Sprinkle with nutmeg and blend in the wine. Pour over the chicken, arranging so the mushrooms are spread evenly. Sprinkle with fresh Parmesan and bake for 45 minutes at 350°F (180°C or gas mark 4).

Eggplant Parmesan

Makes 6 servings.

- 1 onion, peeled and chopped
- 3 tablespoons (42 g) butter
- 2 8-ounce cans (225 g) tomato sauce
- ¼ cup (10 g) chopped basil leaves
- 1½ teaspoons (9 g) garlic salt
- Freshly ground pepper
- 1 large eggplant, peeled and cut into ½-inch (1.3 cm) slices
- 1 egg, beaten
- ½ cup (62 g) all-purpose flour
- Oil for frying
- 1 cup (225 g) mozzarella, sliced
- ½ cup (50 g) freshly grated Parmesan

Sauté the onion in the butter until limp but not browned. Mix in the tomato sauce, basil leaves, garlic salt, and pepper, and bring to a boil. Reduce the heat, and simmer the sauce for 15 minutes. Set aside.

Dip each eggplant slice into the egg and then the flour, and brown on both sides in the hot oil. Drain well.

Place a layer of eggplant in the bottom of a buttered ovenproof dish, topping each slice with mozzarella and some of the sauce. Repeat until all the eggplant is used. Sprinkle the Parmesan over all, and bake for 45 minutes at 350°F (180°C or gas mark 4). Brown under the broiler for 1 or 2 minutes until bubbly on top.

Lemon Pudding

Makes 6 servings.

- 3 eggs, separated
- 2 lemons
- 2 tablespoons (28 g) butter, softened
- ¾ cup (150 g) sugar
- 1 cup (235 ml) milk
- 2 tablespoons (28 g) flour

Beat the egg whites until firm; set aside. Juice the lemons and grate the rind; set aside.

Beat the butter and sugar until creamy. Add the egg yolks one at a time. Add the milk, flour, and lemon juice, and beat well. Fold in the grated lemon rind. Fold in the beaten egg whites. Pour into a 2-quart (30 x 20 cm) baking dish set in a pan of hot water coming halfway up the sides of the baking dish. Bake for 50 to 60 minutes at 350°F (180°C or gas mark 4).

Asparagus in Ham Blankets

Makes 6 servings.

- 1½ cups (290 g) white rice, uncooked, rinsed, and drained
- 3 cups (710 ml) chicken broth or water
- 6 slices cooked ham, cut in half
- 24 asparagus spears, washed and trimmed
- 12 toothpicks
- 1 recipe Velouté sauce (see sidebar, page 58)
- ½ cup (55 g) grated Swiss cheese

Butter a 9 x 13-inch (20 x 30 cm) baking dish. Spread the uncooked rice in the bottom, and slowly pour the hot chicken broth or water over. Place 2 asparagus spears on each piece of ham, and roll up, fastening each with a toothpick. Arrange over the rice, cover tightly with foil, and bake for 45 minutes at 350°F (180°C or gas mark 4). Heat the Velouté sauce and blend in the Swiss cheese. Carefully remove the foil, pour the warmed Velouté mixture over all, and return to the oven for 15 additional minutes. Place under the broiler for an additional 1 or 2 minutes if desired.

Creamy Chocolate Pudding

Makes 6 servings.

- 2 squares unsweetened chocolate
- 3 cups (710 ml) scalded milk
- 1½ cups (75 g) firm, fresh bread crumbs (see sidebar)
- ¾ cup (150 g) sugar
- 3 tablespoons (42 g) melted butter
- 2 eggs, slightly beaten
- 1 teaspoon (15 ml) vanilla extract
- Pinch of salt

Butter a 2-quart (30 x 20 cm) baking dish. Melt the chocolate in the hot milk, and blend until smooth. Add the bread crumbs and let cool. Add the remaining ingredients and mix well. Pour into a buttered dish, and bake for 50 to 60 minutes at 350°F (180°C or gas mark 4), until set.

Sauces

These two sauce recipes can be doubled or tripled and kept in the freezer in small portions, ready for use.

Rich Brown Sauce

- 2 tablespoons (28 g) butter
- 3 tablespoons (30 g) finely chopped onion
- 3 tablespoons (42 g) flour
- 1 cup (235 ml) hot beef broth
- 1 tablespoon (15 ml) Worcestershire sauce
- Salt and pepper to taste.

Cook the onion in the butter in a heavy-bottomed pan or skillet. Add the flour, and continue to cook until it's a dark brown color, about 10 minutes. Add the hot broth and bring to a boil, stirring constantly until thickened, about 1 minute. Blend in the Worcestershire sauce, and taste before adding salt and pepper. Makes about 1 cup (235 ml).

Sauce Velouté

- 2 tablespoons (28 g) butter
- 3 tablespoons (42 g) flour
- 1 cup (235 ml) hot chicken broth
- ½ cup (+120 ml) half-and-half
- Salt and pepper to taste

Melt the butter in a heavy-bottomed pan or skillet. Stir in the flour, and blend well. Cook over medium heat 2 minutes, stirring constantly to keep from burning. Add the hot chicken broth, continuing to stir as sauce thickens. Bring to a boil, lower the heat, and cook for 2 more minutes. Add the half-and-half and blend thoroughly. Taste before seasoning further. Makes about 1½ cups (355 ml).

Bread Crumbs

To make soft bread crumbs, place cut-up pieces of firm bread, crusts trimmed, in a blender or food processor, and process until no large pieces remain. To make buttered bread crumbs, lightly toast slices of firm bread, butter them, and place in broken pieces in a blender or food processor, processing until no large pieces remain.

Eat Raw Foods for Enzymes and Save Energy

Cost savings

About $11 per week, based on several substitutions in a weekly menu, plus savings in cooking costs

Benefits

Saving energy, eating "living" foods

Going raw is nutritious, convenient and easy on your wallet if you rely on your garden for meals. Make the most of enzyme-rich foods by eating them as is (or dehydrating them). *BackHome* contributor Anna Lauer Roy shares how.

Eating living foods enlivens our own life forces. It may be clichéd, but as they say, you are what you eat. Raw or "living" foods are foods that have not been heated above 115°F (46°C). In their raw form, fruits, vegetables, nuts, legumes, and grains contain enzymes that are essential to optimal digestion and absorption of a food's nutrients. Cooking a food kills virtually all of its enzymes. Eating even a few raw-food meals a week can help supply your body with the living enzymes that assist blood cells in all the work they do toward keeping your bodily systems vital.

Most of us eat a lot of raw produce (green salads and fresh summer fruits, for example) in the spring and summer, but as fall and winter approach, it can seem less obvious to eat foods in their living and most energy-packed form. Some people crave warmer, richer fare as cold weather sets in. But there are delicious, fun, and hearty ways to turn fall harvest produce into nourishing and sumptuous meals. In fact, autumn is a prime time to experiment in the art of living foods preparation because of the abundance, variety, and nutritive wealth that we reap from harvests of produce such as winter squashes, pears, and root vegetables in September and October.

In the fall and winter when fresh greens are becoming scarcer, sprouting seeds and grains indoors will keep your family rich in fresh, leafy, and very local greens. Sprouts make beautiful garnishes, sandwich fillings, and snacks. Also, you can make green salads entirely from sprouts, as Sprouted Leafy Greens Salad (below).

The following menu, which serves four, was created to make the most of fall's bounty and to celebrate the richer, earthier flavors that we sometimes have a hunger for at that time of year.

Sprouted Leafy Greens Salad

Makes 4 servings

- 2 cups (100 g) sprouted sunflower seed greens
- 2 cups (100 g) sprouted buckwheat greens
- 2 cups (100 g) sprouted pea greens (or any combination of sprouts or other greens available)
- 4 tablespoons (60 ml) olive oil
- 2 tablespoons (28 ml) apple cider vinegar
- 1 teaspoon honey
- ½ teaspoon ground dry mustard
- ½ teaspoon ground cumin
- Sea salt
- Black pepper

Toss the sprouts in a serving bowl. Whisk together the oil, vinegar, honey, mustard, cumin, and salt and pepper to taste in a separate bowl. Dress the sprouts salad with some or all of the dressing. Garnish with nasturtium petals or other edible flower petals if available.

Note: Sprouted seeds, grains, and nuts are high in lecithin, vitamins C and E, niacin, and much, much more.

Carrot-Ginger Soup

Makes 4 servings

- 5 large carrots, diced
- ⅓ cup (80 ml) extra-virgin olive oil
- ¼ cup (60 g) raw tahini
- 3 tablespoons (45 ml) nama shoyu (raw soy sauce)
- 3 tablespoons (45 ml) apple cider vinegar
- 3 tablespoons (18 g) diced fresh ginger (amount can be varied according to taste)
- 2 teaspoons honey
- 1 clove garlic (amount can be varied according to taste)
- ½ teaspoon ground licorice root or ground aniseed
- Pinch each of cinnamon, cumin, cardamom, turmeric, curry, and (optional) cayenne pepper

Combine all ingredients in a food processor and blend until very smooth. The soup is delicious at this stage, or, if desired, press through a strainer to remove pulp. You can heat the soup very gently if you wish, but be sure not to heat above about 115°F (46°C) to ensure that the living enzymes stay intact. For a quick garnish, make carrot curls with a vegetable peeler.

Note: Raw carrots are an excellent source of vitamin A and potassium; they contain vitamin C, vitamin B_6, thiamine, folic acid, and magnesium. Raw ginger, among its many heroic capacities, has been shown to improve circulation, curb nausea, aid digestion, and be soothing for cold symptoms and sore throats.

Pumpkin Dumplings in Tomato-Sage Sauce

Dumplings

Makes 4 servings

- 3 cups (360 g) diced pumpkin
- 1 cup (100 g) raw walnuts
- ½ cup (118 ml) extra-virgin olive oil
- 1 tablespoon (2 g) dried sage (or use fresh if you have it)
- Sea salt
- Black pepper

Place all the ingredients except salt and pepper in a food processor and process until smooth; add salt and pepper to taste. Scoop teaspoonful of the dough onto the dehydrator sheets, forming half-moon shapes, and flatten them slightly with the back of a spoon. They should be thin enough to dry through to the middle. Dehydrate for about 8 hours, or until the dumplings are crispy and come easily off the dehydrator trays. (Once dried through, they can be kept in the dehydrator at its lowest setting until you're ready to serve them.)

Note: Raw pumpkin is rich in antioxidant carotenoids, especially alpha and beta-carotenes, and it is full of vitamins C, K, and E, as well as magnesium, potassium, and iron.

Tomato-Sage Sauce

- ½ cup (118 ml) olive oil
- 2 large tomatoes
- 1 tablespoon (2 g) dried sage (or fresh)
- Sea salt

Cut the stem base out of the top of each tomato, peel the skin off, and squeeze out the seeds. (You can compost the peels and seeds.) Put all of the ingredients in a blender and blend until smooth. (Optionally: Put the mixture in a bowl and place it on top of the dehydrator to warm for the last hour or so of dehydrating the dumplings.) Top the dumplings with the sauce and serve while still warm. Garnish with more fresh sage if available.

Note: If you dry these in a food dehydrator, which is the preferred method, you will need to allow about 8 hours drying time before serving; if you do not have a dehydrator, see sidebar for oven drying instructions.

Dehydrating Dumplings in Your Oven

Turn your oven to the very lowest setting. Grease baking sheets with olive or sunflower oil and scoop teaspoonfuls of the dough onto the baking sheets; flatten them slightly with the back of a spoon. They should be thin enough to dry through to the middle. Put the dumplings into the oven and leave the door ajar. This will keep some air flowing through the oven and will ensure that the oven temperature doesn't rise too high. Monitor the dumplings: When the tops seem dry (after an hour or two), flip them over with a spatula and continue to dry. In the oven, they should take anywhere from 2 to 4 hours to dry. When they look dry on the outside, test one to see if it is dried through.

Pickle Foods with Appeal

Cost savings

Averages about $1.68 per quart (1 L)

Benefits

Healthful and fresh pickled goods for eating or putting up

Pickling is one of the oldest methods of preserving foods, known to have been practiced at least 4,000 years ago. While there are many references to pickling throughout recorded history, one of the most interesting to North and South Americans is that Amerigo Vespucci (the Americas are named after him) was a major supplier of pickled foods to oceangoing ships; the pickles helped protect sailors from scurvy.

Here, contributor Patricia Rutherford offers a few of her favorite pickle and pickled food recipes to confirm that a variety of foods can be delicious when pickled.

> **Note:** Since the latest canning practices advise subjecting pickles and relishes to a boiling-water bath—15 minutes for pint (475 ml) jars, 20 minutes for quarts (946 ml)—we recommend you add this schedule to the five recipes that follow.

Ginger Pickles

These pickles are also known as old-fashioned 14-day pickles. They take longer to make than others, but they're worth the time. You'll need a 4- to 5-gallon (17.2 to 18.9 L) soup pot as well as a large crock for the process, and 15 pint (475 ml) jars for canning.

- 7 pounds (3.2 kg) 4- to 5-inch (10 to 13 cm) pickling cucumbers (about 36)
- 1 cup (225 g) pickling salt (not iodized)
- Green grape leaves (see recipe)
- 6 cups (1.4 L) vinegar
- 4 ounces (115 g) ground ginger
- 4 pounds (1.8 kg) sugar, plus 2 cups (400 g)
- Stick cinnamon (see recipe)
- Celery seed (see recipe)

For crisper pickles, soak the cucumbers in ice water for 4 to 5 hours. Then make a brine of 1 cup (300 g) salt to 1 gallon (3.8 L) water in the soup pot, and soak the cucumbers in it for 7 days, keeping the cukes under the brine with a weighted plate. After 7 days, remove from brine, wash, and cover with fresh water. Keep for three days, changing water every day.

Empty the pot, then into it place alternating layers of cucumbers and grape leaves until the cucumbers are used up. Add 2 cups (475 ml) of vinegar and enough cold water to cover. Bring to a full rolling boil, then pour through a colander, discarding the liquid.

Cut the cucumbers into 1-inch (2.5 cm) -thick pieces. In the pot, bring 2 gallons (7.6 L) of water to a boil and add the ground ginger. After stirring, add the cucumbers. Let stand overnight, stirring occasionally when convenient. In the morning, drain the cucumbers and place in a crock. In a pot, bring 4 cups (950 ml) of vinegar and 4 pounds (1.8 kg) of sugar to a boil. Sprinkle stick cinnamon and celery seed over the pickles in the crock, and mix throughout. Pour the hot vinegar-sugar syrup over. Let stand overnight.

In the morning, drain the syrup into a pot, add 1 cup (200 g) of sugar, bring to a boil, and pour back over the cucumbers in the crock. Let stand overnight, and repeat the next day, adding 1 more cup (200 g) of sugar, boiling, the returning to the crock, and standing overnight.

On the final day, have sterilized pint (475 ml) jars and lids ready. Put the cucumbers and syrup into the pot, bring to a boil, and seal in the sterilized jars. Yields 15 pints.

Dill Pickles

- 12 pints (5.5 kg) 3-inch (7.6 cm) (straight) cucumbers
- 6 teaspoons (about 12 g) pickling spice
- 12 dill flower heads
- 2 quarts (1.9 L) vinegar
- 1 quart (946 ml) water
- 1 cup (300 g) coarse salt

Scrub the cucumbers and soak them in cold water overnight. The next day, sterilize 6-quart (5.7 L) canning jars and lids. Place a teaspoon of pickling spice and 2 dill flower heads into each jar. Dry the cucumbers and pack them closely into the jars.

Bring the vinegar, water, and salt to a boil, pour it over the cucumbers in the jars, and seal them. It's best to wait at least five days for the pickles to ripen before serving. Yields 6 quarts.

Apple and Onion Relish

- 20 green tomatoes
- 8 apples
- 4 red bell peppers
- 8 onions
- 3 cups (600 g) sugar
- 3 cups (720 ml) vinegar
- 3 tablespoons (54 g) salt
- 1 teaspoon (2.3 g) ground cinnamon
- 1 teaspoon (2 g) celery seed

Have ready 9 sterilized pint (475 ml) jars with lids. Finely dice the first four ingredients. Combine in a pot and stir in the remaining ingredients. Bring to a boil, pack into the jars, and seal. Yields 9 pints.

Pepper Relish

- 24 green bell peppers
- 12 onions
- 1 cup (120 g) finely diced celery
- 2 cups (480 ml) vinegar
- 2½ cups (500 g) sugar
- 3 tablespoons (54 g) salt

Bring about 1 quart (946 ml) of water to a boil. Meanwhile, finely dice the peppers and onions, and place them in a pot with the celery. Pour boiling water over to cover; let sit 20 minutes. Drain off the water and add the vinegar, sugar, and salt to the vegetables in the pot. Cook for 20 to 30 minutes. Sterilize the pint (475 ml) jars and lids. When the relish is ready, pack it into the jars and seal them. Yields 6 to 8 pints (depending on size of vegetables).

Pickled Peaches

- 1 peck (16 pounds, or 7.3 kg) clingstone peaches
- 3 cloves per peach, heads removed
- 7 cups (1,400 g) sugar
- 4 cups (960 ml) vinegar

Sterilize four quart (946 ml) jars and lids. Pour boiling water over the peaches, drain immediately, and peel. You may want to spray the peeled peaches with a light solution of lemon juice and water to keep them from turning dark.

Pierce each peach with 3 cloves (the heads are removed to keep the syrup from darkening). In a pot, stir together the sugar and vinegar until the sugar is dissolved, while bringing to a boil. Add the peaches and simmer for about 10 minutes or until tender. Transfer the peaches from the pot into the jars, and pour the boiling syrup over. Seal at once. Yields 4 quarts.

Preserve and Economize with Water-Bath Canning

Cost savings

Putting up your own garden produce can cost less than 10 percent of what it costs to purchase canned foods

Benefits

Knowing the food you preserve is safe and fresh, and made with recipes of your choosing

Before supermarkets were stocked with canned groceries, people relied on water-bath canning to preserve their summer's bounty. Today, it's a great way to save money and extend the life of your garden goodies. Here, *BackHome* contributor Kim Erikson explains the art of water-bath canning.

There's really no mystery to water-bath canning. Acidic food is placed in sealed glass containers and submerged in boiling water for a specified amount of time. Before you start, you'll need to invest in some supplies that you'll use over again: a water-bath canner and good, quality jars.

Often made of graniteware, a water-bath canner looks like a giant soup pot and comes in 21-quart (19.8 L) and 33-quart (31.2 L) sizes. It should come with a rack for holding the jars during processing. Many housewares stores stock them or can order one for you. You'll also need good quality Mason jars such as Ball or Kerr, which you can purchase at most grocery stores. Avoid using Aunt Heddy's antique canning jars, and never use commercial food jars—the kind store-bought mayonnaise or pickles come in. They are too thin to take the heat and cooling and can crack and chip easily. Since canning jars can be reused, consider your purchase a worthwhile, long-term investment.

You'll also need to purchase screw bands and lids that fit your jars. Screw bands can be reused as long as they haven't rusted. The lids, however, are used only once because the sealing compound can't seal adequately after its initial use. Once you've eaten your home-canned goodies, throw the lid away.

EASY AS 1, 2, 3

The two keys to successful canning are planning and careful attention to the processing instructions and the recipe. First, gather your supplies. You'll need:

- assorted bowls
- measuring spoons and cups
- knives, tongs, a slotted spoon, a ladle
- colander
- dish towels
- jar lifter
- wide-mouthed funnel
- jars (half-pint [235 ml], pint [475 ml], and quart [946 ml] sizes)

STEP 1: Prepare the canner. Place the rack inside the canner and fill it with enough water to cover the jars by 2 inches (5 cm). Set the canner over high heat and replace the lid. It takes a while for the water to boil, so start early. Start heating a large pot of water in which to sterilize the jars.

STEP 2: Sterilize the jars. Carefully check the jars, discarding any with nicks, chips, and cracks, especially around the rim. Wash usable jars in hot, soapy water or run them through the dishwasher. Sterilize the jars by placing them in the pot of boiling water for at least 15 minutes. Once they're sterilized, they can be held in the hot water until you're ready to fill them.

STEP 3: Wash lids. Wash the lids and screw bands in hot, soapy water. Place the lids in a shallow bowl and pour boiling water over them to soften the sealing compound.

STEP 4: Prepare your food. Select fresh, high-quality produce, preferably picking it the same day you process it. Wash the food, then prepare it according to the recipe. Unless otherwise instructed, cut the food into pieces approximately the same size and shape. Not only does the food process more uniformly, the finished product looks more attractive.

STEP 5: Fill the jars. There are two methods for filling jars: cold pack and hot pack. Cold pack, also called "raw pack," is used for some fruits and most pickles. Raw food is placed in the jars and then boiling liquid is poured over it. Cold-packed food usually requires a longer processing period. Hot pack requires cooking the food before it's placed in the jars. Jams, jellies, sauces, and fruit purees use the hot-pack method.

A word about headspace: Leave some room between the top of the food and the jar's lid (usually about 1/2 inch, or 1.25 cm) to allow for expansion of the jar's contents while processing.

With your tongs, carefully lift the sterilized jars out of the scalding water and set them on a clean dish towel. **For cold pack:** Place the food in the jar, arranging it as neatly as possible. Using a wide-mouthed funnel, quickly pour the hot liquid over the raw food, remembering to leave 1/2 inch (1.25 cm) of headspace.

For hot pack: insert the wide-mouthed funnel into the first jar and fill it with the hot, cooked mixture. Repeat until all the jars have been filled. Quickly place a hot lid on top of each jar, wiping the rim with a clean, damp sponge to remove spilled food before securing the lid with a screw band.

STEP 6: Submerge jars. Using the jar lifter, place the jars into the boiling water-bath canner, making sure the jars are upright. Water should completely cover the jars. Replace the canner's lid and process for the amount of time specified in your recipe.

STEP 7: Allow jars to cool. When the food has finished processing, transfer the jars from the canner to a heatproof surface. Let the jars sit until they're completely cool to the touch. You should hear a loud popping sound as the lids complete their sealing process. Then, test the seal by pressing down on the center of each lid. There should be no give. If the lid moves up and down at all, store the jar in the refrigerator. Tighten the screw bands and label each jar with the contents and date of processing. Store your finished product in a cool, dry place.

Now that you have the basics, you can try an applesauce recipe. Other books to consult include the *Ball Blue Book: The Guide to Home Canning and Freezing* (Alltrista Corporation) and *The Big Book of Preserving the Harvest* by Carol W. Costenbader (Storey Publishing).

Applesauce

Yield: 4 pints

- 20 large, firm apples
- 4 cups (946 ml) water
- 1 tablespoon (14 ml) lemon juice
- Sugar
- 1 teaspoon ground cinnamon

Peel, quarter, and core the apples, removing bruised spots. As each apple is prepared, place in a bowl of cold water to which 1 tablespoon of lemon juice has been added. When all the apples have been prepared, drain and rinse under fresh water. Place in a large pot with the 4 cups (946 ml) of water and bring to a boil. Reduce the heat, and simmer until tender. Mash apples until smooth. Add sugar to taste along with the cinnamon. Return the mixture to a boil, stirring constantly. Ladle into hot, sterilized pint (475 ml) jars, leaving a ½-inch (1.3 cm) headspace. Seal. Process for 25 minutes.

Observe Canning Safety Rules

Cost savings

The savings in cost are in the canning itself, but these precautions still have value

Benefits

Reducing the risk of burn injuries or possible illness from spoiled foods

Home canning has become extremely popular as people realize the cost and health benefits of preserving food from their own gardens. But canning safety is an important matter that cannot be rushed through or ignored. Here are a dozen tips for maintaining safe canning practices when you put up your harvest.

1. Foods that are high in acid, such as fruits and tomatoes, can be canned in a boiling water bath canner. Low-acid foods, such as most vegetables, will need to be processed in a pressure cooker.
2. Extremes hot and cold temperatures will crack jars. Put hot food into hot jars, and do not put hot jars onto cold surfaces.
3. Remove air bubbles with a rubber spatula or a wooden spoon; air bubbles encourage spoilage.
4. Always check a jar of food before serving it. It should look and smell normal and the jar should be sealed firmly. Simmer all low-acid foods for 15 minutes before serving. If there are any "off" odors, throw the food out.
5. Using a jar with a tiny chip in the top or a small crack in the side will result in either a broken jar or an incomplete seal. Check all jars carefully before you fill them.
6. Adjust your processing time according to altitude. Most charts are calculated to altitudes of 1,000 feet (305 m) or less. Increase the processing time by 5 minutes for altitudes of 1,000 to 3,000 feet (305 to 900 m), 10 minutes for altitudes of 3,000 to 6,000 feet (0.9 to 1.8 km), and 15 minutes for altitudes of 6,000 to 8,000 feet (1.8 to 2.4 km).
7. Add more boiling water to the canner if water does not cover the tops of the jars by at least 1 inch (2.5 cm).
8. Always use the wire rack of your canning pot, as the boiling water must circulate under, around, and over the jars. The wire rack will prevent overheating and cracking of the jar bottoms and keep the jars from knocking together.
9. Don't invert your jars when removing them from the canner. It can interfere with the sealing process and hinder cooling.
10. Always check the seal as you store the jars. Each jar lid should be indented in the center, without giving as you press down with a finger. If it makes a noise or if it can be pushed down, it is not sealed and must either be reprocessed with a new lid or be eaten immediately.
11. Don't attempt to double canning-manual recipes or otherwise alter them. You may run into problems, especially if you are a beginner.
12. Always be careful of steam escaping the canner and the hot jars—it can burn you. Lift the canner lid away from you to let the steam escape safely.

CHAPTER 2

LAUNDRY ROOM AND LINEN CLOSET

A fair amount of your household budget can go down the drain with the cost of laundry supplies. Keeping clothes and linens clean requires valuable resources—water and energy—and even a month of inefficiency can take a toll.

In this chapter, you'll discover how to approach laundry chores sensibly, and how to replace costly laundry products with low-cost, effective alternatives. You'll also gain energy-saving tips and ideas that are pocketbook- and planet-friendly.

Know the Difference Between Soap and Detergent and Get Your Clothes Cleaner

Cost savings

Between $0.90 and $2.05 per 28-load product if chosen correctly

Benefits

An understanding of what gets laundry clean without causing environmental damage

Doing laundry is more complicated than it seems with the range of products available to get the job done. To get the dirt, Richard Freudenberger, technical editor at *BackHome* magazine, got in touch with the people at Seventh Generation, the Vermont household and personal-care products company that has been a pioneer of environmental and corporate responsibility since its founding in 1988.

Martin H. Wolf, Seventh Generation's director of product and environmental technology (and also the 7Gen blogger "scienceman"), explained that laundry products have gone through a litany of change over the years. So what *is* the difference between soap, detergent, and some of the other cleaning products you might pour into your washing machine on a weekly basis?

THE SOAP EVOLUTION

Soap has been around for a long time. Its basic ingredients are the sodium salts of fatty acids, drawn from animal fats or plant oils and cooked up with lye, an alkaline caustic soda or potash. This works fine if washing is done in soft water. But most U.S. households have some level of hard water, which combines with soap to form a lime you know as "soap scum."

To resolve the problem, chemists developed detergents, which allow us to get clothes clean in hard water, thanks to the work of several chemicals. Detergents contain surface active agents (surfactants) that reduce water's surface tension so the detergent can penetrate fabric better. Detergents also contain agents that bind water hardness minerals (called builders) and suspend soil particles so they can be rinsed away. The ultimate result: no soap scum and clean clothes.

THE MAIN INGREDIENTS

Building detergents from scratch has allowed manufacturers to custom-design their products for consumers. Consider how older laundry detergents substituted phosphates for sodium carbonate and citric acid builders. (This imparts alkalinity to water and "softens" its hardness.) The bad news: Phosphates find their way from household drains to our streams and lakes and become nutrients for algae, which respond with a reproductive bloom, die off, then themselves become nutrients for bacteria, which deplete oxygen from the water. The result is hypoxic zones and fish kills. Though laundry phosphates were banned years ago, an industry-voluntary ban on phosphates in dishwasher detergent only came about in July 2010.

Next, we have surfactants: a costly part of the laundry product. There are several types based on function. The most common surfactants are linear alkylbenzene sodium sulfonates (LAS). These are anionic surfactants in that the surface-active part is a negatively charged ion in water, which reacts with the positively charged water hardness ions of calcium and magnesium. The synthetic LAS do not biodegrade all that well, and do so rather slowly.

Additionally, there are other ingredients added to detergents to impart desirable characteristics for consumers. Besides fragrances, colorants, opacifiers, and suds stabilizers, which are sensory and convenience agents, there may be enzymes that break down the complex proteins found in blood and grass, optical brighteners (dyes that adhere to fabric and emit a visible blue light), and antiredeposition agents, special dispersion polymers made from cellulose that keep soil from redepositing on clean fabrics.

So, there *is* a reason why detergents will do a better laundering job than soap flakes and powders. But the chemical aspect of laundry cleaning is only part of the story.

THE COLD, HARD TRUTH

Want to get clothes cleaner? Mind the water temperature. The effectiveness of soaps and detergents diminishes as water temperature decreases. Below about 60°F (16°C), the cleaning power of any detergent is noticeably reduced, so more detergent may be needed to get clothes clean. Hot water, above 125°F (52°C), should be avoided, if only because it's a costly use of energy. The only exception might be in washing sickbed linens or dirty diapers. Lukewarm water or cold water, when used in accordance with instructions on the label, will work suitably well while saving a significant amount of money over time.

Also consider the efficiency of your washing machine. Generally, modern front-loading (horizontal axis) washers are more efficient than top-loading styles, and features such as optimized load cycles, high-speed spin, and internal heaters are worth looking into if you are in the market for a new machine.

Make Your Own Laundry Soap for Pennies a Load

Cost savings

About 13 cents per laundry load

Benefits

Inexpensive laundry soap that works well and deodorizes, too

If you do a lot of laundry—heck, even if you *don't* do a lot of laundry—you'll probably find this recipe for homemade soap a real money-saver. It costs less than 1½ cents per load and takes about 15 minutes (plus setting time) to make two gallons (7.5 L).

The key ingredients are borax (sodium tetraborate) is an alkaline mineral that comes in powder form and is also an effective deodorizer. Washing soda (sodium carbonate) is a soda ash with a fairly high pH or alkaline level. It is an excellent solvent for grease and oils, and it has deodorant qualities as well. Everything you need (except for the essential oils) should be available on your grocer's shelves and are safe to use with laundry.

Laundry Soap

- ⅓ to ½ bar of Fels-Naptha or Kirk's CoCo Castile soap
- 4 cups (940 ml) water
- ½ cup (154 g) washing soda
- ½ cup (154 g) borax powder
- 1 tablespoon (15 ml) essential oil, such as pine or peppermint

Grate the soap bar and put it in a saucepan with the water. Heat the mixture until the soap dissolves. Add the washing soda, borax powder, and essential oil. Stir until all ingredients are dissolved. Allow the mixture to sit for 5 minutes over the heat; stir occasionally. Remove from the heat and allow to cool for 5 minutes.

Fill two 1-gallon (3.8 L) jugs halfway with hot water. Pour the soap mixture in equal amounts into each jug. Shake well, and then completely fill each jug with warm water. Cap the jugs and shake well. Set aside for 24 hours while the mixture sets up. Application: ½ to ⅔ cup (120 to 155 ml) per wash load.

Premix these small amounts to make 5 gallons (19 L) at a time; 2 gallons (7.5 L) is good for about 50 laundry loads.

Discover the Bonus of Borax

Cost savings

Between $0.70 and $1.66 per three-load laundry week, depending on which cleaning product you are replacing

Benefits

Using a single product to do several laundry chores

Though borax may have been a household word to our grandmothers, today the powdery white compound often remains a stranger to our pantries and laundry rooms. That's a real shame, because a little bit of this household helper can go a long way toward extending your soap supply, freshening and brightening up soiled clothes, and aiding in general cleanup around the home.

Technically, borax is sodium borate (or sodium tetraborate), a fairly benign compound of sodium, boron, and oxygen that does not contain phosphates or chlorine. It's sold in stores under the Dial Corporation brand name 20 Mule Team Borax, though other options exist. Note that borax is *not* the same as washing soda, which is sodium carbonate, sometimes called soda ash. (Arm & Hammer Super Washing Soda is often used to soften hard water, though it's also a laundry booster.)

Here are some ways to use versatile borax around the house.

General Cleaning: For everyday freshening and cleaning of countertops, tile, and other surfaces, a mixture of 1/4 cup (60 g) of borax, a half capful of mild dish soap, and 1 gallon (3.8 L) of warm water. For spot jobs, make a paste using borax and a little water. Use a no-scratch sponge or an old washcloth on sensitive surfaces such as tile or metal cabinet trim.

Laundry Booster: This was the original use for borax, which was mixed with soap flakes to hand-launder clothing and fabrics. It brightens whites and colors. Use 1/2 cup (110 g) with warm water in the wash cycle for best results.

Laundry Pre-Soak: 1/2 cup (110 g) of borax in a normal-sized laundry load will help break down stubborn stains that would otherwise leave a residue. Allowed to soak for a half-hour or so before washing, it works well on blood, food, fruit, coffee, tea, rust, and mildew. It is not so effective on grass, lipstick, or grease stains.

Deodorizer: 1/2 cup (110 g) of borax (or baking soda) in every load of laundry helps neutralize odors. Usually, a single wash cycle will remove the odor; if not, you may have to pre-soak.

Cloth Diaper Soak: A stronger solution of borax (1 cup [225 g] to 5 gallons [19 L] of warm water) makes a good presoak for soiled cloth diapers. Soak the diapers for 12 hours in a covered container, then wash immediately after.

Water Softener: Use 1/2 cup (110 g) of borax in the washing machine to help soften hard water without leaving a dull carbonate deposit on clothing, which washing soda may do. (Soft water contains fewer minerals and does not leave mineral deposits on clothes, therefore less soap is needed to brighten fabric.)

Use the Sun and Save on Utilities

Cost savings

Between $46 and $97 per year, depending on your utility costs

Benefits

Naturally air freshened clothes, dried at no cost

Rising energy costs naturally boost the price of doing laundry at home. One way to save energy is to reconsider the conventional clothes dryer. The average gas dryer uses about 17,000 BTUs of heat throughout its roughly 45-minute cycle. (A BTU, or British Thermal Unit, is a measure of the amount of heat it takes to raise 1 pound (475 ml) of water 1 degree Fahrenheit [0.5°C.) Converted to dollars, that amounts to 21 to 26 cents, depending on the cost of natural gas where you live. Electric dryers are even more expensive to operate: a 40-minute cycle absorbs 3.3 kilowatt-hours of electricity, which ranges from 33 to 69 cents, depending on local utility charges.

Now, consider the number of times you use the dryer each week, and multiply that by 52. Your annual bill can reach $100—and that's just for drying your clothes. If you hang-dry your clothes on a line instead, you capture the sun's all-natural BTUs: about 16,700 of heat energy for every 10 square feet (3 square m) of the earth's surface.

A clothesline doesn't have to be fancy. You can sink two 8-foot (2.4 m) 4 x 4's in the ground 18 or 20 feet (5 to 6 m) apart, string a line between them, get some clothespins, and you're in business. If you're a stickler for efficiency, you can purchase a single-pole fold-up umbrella model for as little as $56 and hang rows and rows of laundry in one session.

True, your clothes will take most of a day to dry, but you reap the benefit of a sun-soaked, full-spectrum airing that not only destroys mold spores but is less abrasive to your fabrics. And you can't beat that natural-sun smell.

Those who can't hang dry their clothing can use freestanding racks, wall-mounted scissor frames, and even retractable laundry cords to dry laundry inside.

Join the Cloth Diaper Revival

Cost savings

For a single child, the savings is at least $385 annually, probably more, based on a cost of $735 per year for disposable diapers. Because diapers can be reused for subsequent children, the cost savings is even greater for larger families.

Benefits

Cloth diapers are cleaner, cheaper, and just as reliable as disposable diapers.

Even 20 years ago, 18 billion disposable diapers were thrown into American landfills. Thirty percent of current disposable diapers are made of plastic, which is not readily biodegradable. What's more, 3.4 billion gallons of oil and more than 250,000 trees a year are used to manufacture disposable diapers.

Disposable diapers consume a hunk of the household budget for families. The average family spends at least $735 a year on diapers. Considering some children wear diapers until they're age 4 (and older at nighttime), that number can double, triple, quadruple, or more before the child is potty-trained.

With those thoughts in mind, it wasn't difficult for Susan Dabuiskis-Hunter, a new mother and contributor to *BackHome* magazine, to make a commitment to cloth diapers.

I folded the diaper into thirds, fitted it snugly into a Velcro diaper wrap I had purchased weeks before, took a deep breath, and, there, my baby daughter was wearing a cloth diaper. Being a new mother, I worried about what I was doing. Without pins would it stay in place? Would it leak? Was the diaper cover too tight? Too loose? Would she get a rash? Was I somehow depriving her of a benefit of modern technology?

Other questions involved sanitation. My husband was concerned about the odor of diapers waiting to be washed. I wondered about laundering some 6,000 diapers over the upcoming years. Could I get them clean enough? We live in an apartment building with a communal laundry, and I worried that the neighbors would balk at the thought of using a washer after it had held a load of dirty diapers.

I can happily report these worries dissipated once I started using the diapers. Cloth diapering is simple and rewarding (my husband agrees). Our baby is content, her bottom is healthy and soft, bills for disposable diapers are nonexistent, and the neighbors have not complained.

Before our daughter was born, I went online to research cloth diapering, and I was astounded at just how many users and supporters there are. The relatively recent creation of safe and convenient pinless diapers and Velcro-fastened covers have brought things a long way from the nasty pins and the plastic pants of yesteryear. Another consideration is comfort. Surveys of older incontinent people who had used both cloth and disposable diapers for extended periods of time had unanimously voiced their preference for the comfort of wearing cloth next to their skin.

DISPOSABLE DANGERS

Comfort aside, I wondered about the gel in disposables that could contain so much liquid without appearing wet. No research is available on what the long-term effects of the chemicals used to absorb urine may have on future fertility or cancer rates after 2 to 4 years next to children's delicate skin. By the time a child reaches the age of 2 and a half, he or she will have spent more than 20,000 hours next to these super-absorbent chemicals.

One of these is sodium polyacrylate, whose purpose is to pull fluid away from the baby's skin, absorb this fluid and hold it in the diaper. Sodium polyacrylate has been linked to toxic shock syndrome, allergic reactions, and death if ingested even in small amounts. Cats have died inhaling this substance. Pediatric journals cite reports of this chemical leaking from diapers and sticking to babies' genitals.

The Environmental Protection Agency has stated that dioxin, a byproduct of the bleaching process of the paper used to make disposable diapers, is the most toxic of all cancer-linked chemicals. Even in the smallest detectable quantities, dioxin has been known to cause liver disease, immune system suppression, and genetic damage in lab animals.

The Food and Drug Administration has received reports of diaper fragrances that have caused headaches, dizziness, and skin rashes. According to the *Journal of Pediatrics*, 54 percent of one-month-old babies using disposable diapers had rashes, and 16 percent had severe rashes. Other reasons for rashes attributed to disposables are lack of air, higher temperatures due to plastic retaining body heat, and less frequent changes because diapers feel dry even when wet.

Problems associated with disposable diaper use, as reported to the Consumer Protection Agency, include chemical burns, noxious chemical odors, babies pulling diapers apart and putting pieces of plastic into their noses and mouths, babies choking on tab papers and liners, and ink staining the skin with dyes known to damage the central nervous system, kidneys, and liver. Most recently, emissions from disposables were found to cause asthma attacks and respiratory problems.

EASY CARE

Cloth diapers are convenient if you consolidate your washing. When washed in hot water with regular detergent, cloth diapers do not need rinsing beforehand. Simply throw wet and soiled diapers into a lined and covered pail, and wash them every two or three days.

During diaper changes, I clean my baby's skin with plain warm water and baby washcloths. The soiled cloths go into the pail with the diapers. I don't use baby wipes, because a doctor suggested I try washing my hands only with wipes all day to sense what a baby feels like cleaned that way.

After cleaning my baby's bottom, I dry her skin with a hair dryer set very low but still warm. She loves it! I have six outer covers, which I switch with each diaper change. These I either air-dry, hand-wash, or launder with the diapers, depending on how soiled they are.

All this is easier, of course, because I stay at home with my daughter. If I couldn't, though, I think I would still find a way to use cloth diapers. I'm saving money, helping the environment and doing something healthy for my baby.

USING A CLOTH DIAPER

Start with one diaper (a) and one cloth diaper cover. Place the diaper on top of the cover, the inner sides of both facing up (b). Fold the bottom left and right corners to the middle, creating a Y shape (c). Bring it up between the baby's legs (d) and fold it to adjust it to your baby (e), making sure the cover extends above the diaper. Fasten it securely.

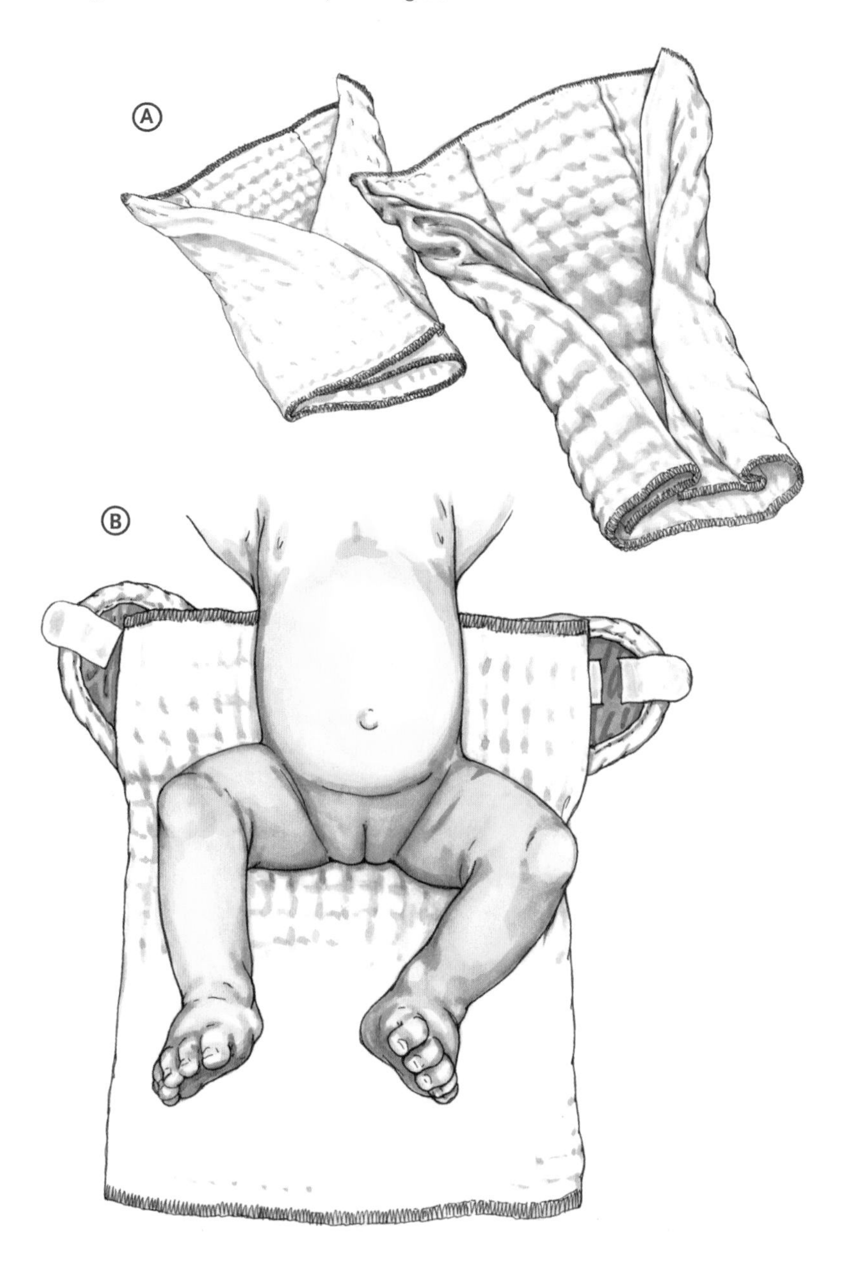

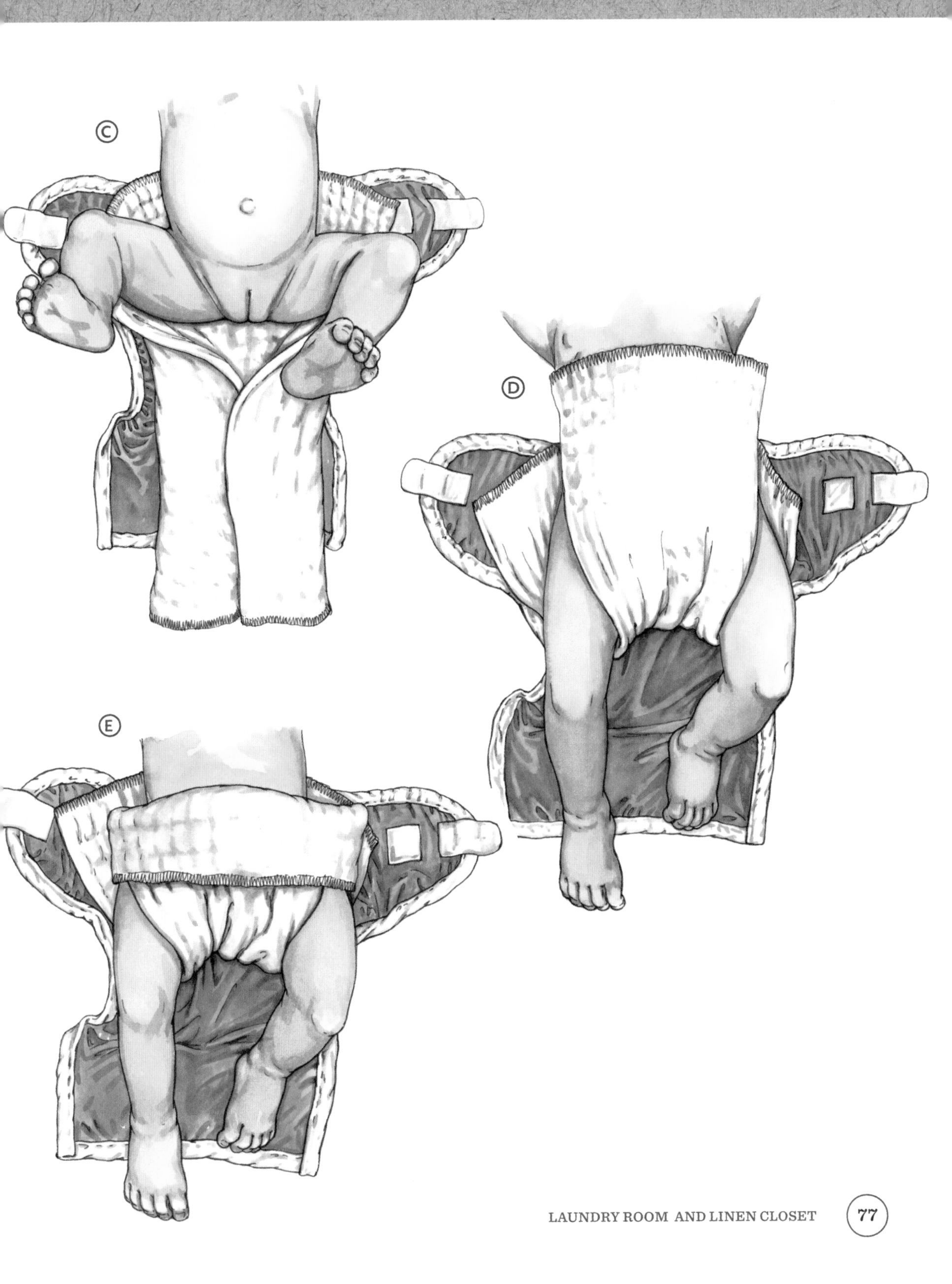
C
D
E

Find a Dozen Uses for Old Sheets and Linens

Cost savings

Between $4 and $25, depending on the item

Benefits

Repurposing fabric for uses you may not have thought of.

Give your old sheets and linens new life. The *BackHome* staff brainstormed a dozen uses for old bedding:

1. Old linens make nice picnic spreads or tablecloths to cover outdoor park tables when enjoying an open-air meal.
2. Sheets that aren't torn or terribly worn can make decorative—and nicely coordinated—window curtains. Curtains are an easy sewing project because they only require simple hems and can be hung through sewn sleeves or from sewn-on tabs.
3. Old sheets, and particularly fitted sheets, make ideal frost protection for fruit trees during those early spring frosts.
4. Cut sheets into smaller sections and make utility bags out of them, as for gathering garden produce or holding clothespins for the clothesline. Simply-stitched seams and handles make this an easy job. Sew in a sleeve and make drawstring handles if you wish.
5. Cut clean linens into broad strips and use them for baby wipes, dishcloths, dustcloths, stain-wiping cloths, and so forth. Cut them into narrow strips and they can be stored in a first-aid kit for use in holding bandages, making tourniquets, and tying. Large clean sections of folded cloth make essential compresses in emergency bleeding situations.
6. Make doll clothing from the most colorful print sheets. Hems and piping from pillowcases are ready-made for doll-clothes detailing.
7. For all you quilters, flat sheets still in decent condition can be used as quilt backing. Patterned cloth can be cut into squares for the face side.
8. Tear sheets into narrow strips and use a large crochet hook to make them into rugs. Thick braided rugs can also be made using heavier flannel sheets.
9. Cut colorful sheets into squares with pinking shears and make preserve covers for the canned goodies you plan to give as gifts.
10. Full sheets make great pet beds or covers for your car seats when traveling with your pooch.
11. Use old linens and sheets to teach youngsters how to sew. This applies to hand-sewing as well as machine sewing, and mistakes waste nothing—everything is a learning experience.
12. In the winter, cover your car's windshield with a flat sheet and tuck the ends into the seam between the doors and windshield pillars. In the morning, you can peel off the fabric and be on your way with no scraping or defrosting.

Try a Natural (and Safer) Alternative to Bleach

Cost savings

Little if any monetary savings, but easier on fabrics

Benefits

Hydrogen peroxide is much safer than bleach and is environmentally benign

Ordinary chlorine bleach is the most common source of household poisoning, explaining why parents lock up their sink cabinets to avoid potential disaster. But simply using bleach can pose long-term health and environmental risks.

Chlorine bleach is sodium hypochlorite, a toxic chlorine salt that reacts with a variety of common organic chemicals to form organochloride compounds. (You've probably heard of volatile organic compounds, or VOCs.) Many of these compounds are suspected carcinogens. When washed into the wastewater stream through laundry and sink discharges, chlorine bleaches connect with elements in waterways and soil and react in unpredictable ways. This is a potential public safety risk.

The good news is, there are safe alternatives to chlorine bleach that are fully capable of managing our household cleaning and disinfecting chores. Enter: hydrogen peroxide. Yes, it's a type of bleach. But it's safe to use in laundry, in the kitchen, and around the home for cleaning.

LAUNDRY ROOM: Newer oxidizing laundry detergents often include hydrogen peroxide in their ingredients because it whitens fabrics without the use of chlorine. Over-the-counter hydrogen peroxide (3 percent) is inexpensive in volume and needs to be stored in its opaque container away from light and heat.

Rather than using household bleach when laundering whites, simply pour ½ to ¾ cup (120 to 175 ml) of hydrogen peroxide into soapy water before adding your clothing and linens. Don't pour it directly onto fabric or it may fade or spot the cloth.

The kitchen: Mix one part peroxide with one part water and use the solution to clean countertops and appliance surfaces. You can use it in a spray bottle as well, but finish what you make; the mixture decomposes when exposed to light. Hydrogen peroxide can also be used to disinfect cutting boards after rinsing them off, but you should use it full strength as it comes out of the bottle for this purpose.

On surfaces: To treat mold and mildew, mix one part hydrogen peroxide with two parts water and use it from a spray bottle to neutralize the mold spores. Spray lightly; do not soak. Leave it on the surface to dry for maximum effectiveness. Stronger half-and-half mixes can be used directly on affected surfaces and wiped off, but be aware of peroxide's bleaching characteristics on fabrics.

CHAPTER 3

BATHROOMS

If there one place in the home that you want to keep clean and fresh, it's the bathroom. But this space is so much more than a place to bathe: it's where we go to rejuvenate and relax. The bathroom should offer a healthy environment to unwind. This chapter provides money-saving cleaning strategies that save time and resources—and we treat you with some recipes for homemade body care products.

Make Your Own Housecleaners

Cost savings

Between $2 and $5 or more, depending on the product being replaced

Benefits

Cost savings while removing harmful chemicals from your household

Over the years, *BackHome* magazine has published a variety of articles and letters about the hazards of commercial household cleaners. Aside from the toxicity of commercial cleaners, they are expensive to buy. Fortunately, some basic grocery items can work double-time to spruce up your home-sweet-home.

SPOT CLEANERS

Simple ingredients can work wonders: salt, lemon juice, club soda, and cream of tartar. Here are some "clean" mixes.

Rust stains: Remove rust stains from real porcelain sinks and tubs by dampening a sponge or soft cloth with cream of tartar (a mildly acid salt) and rub out the stain.

Red wine stains: Dab fabric with club soda, an alkaline. (Don't rub; it may spread the stain.)

Carpets and fabric stains: Sprinkle cornstarch on stains and let sit for 10 to 15 minutes. Vacuum excess powder.

Perspiration stains: Pre-treat sweat stains by mixing 1/4 cup (72 g) of table salt in 1 quart (946 ml) of hot water. Apply with a sponge, then rinse.

Paint and rust stains: Remove these tough stains from fabric with a lemon-juice daub. Just be sure to rinse and wash immediately afterward.

Neutralize pet "accidents": Mix one part distilled white vinegar with three parts water. Dab solution on to fabrics or carpet.

GLASS CLEANER

Graphic artists and illustrators have known this trick for years. Possibly the finest streak-free glass cleaner you can use is simply a 25:75 mix of distilled white vinegar and water, soaked into white (not colored) newsprint. Wad up the newspaper sheet, wet it, and rub the glass clean. Dry with a fresh sheet of newsprint. If there is a film on the glass, add a drop or two of liquid soap for the initial wash; thereafter you can use just the vinegar mix.

LEATHER STAIN TREATMENT

Damaged shoes, leather clothing, furniture, and antique desktops can benefit from vegetable-based glycerine. It absorbs moisture, works as a lubricant and cleans surfaces. It performs well on grease and oils. Simply apply glycerine to the stain (don't use on suede) and let it soak in. Wipe off with a cotton cloth, then dab away residue with warm water and a mild soap.

TOILET BOWL CLEANER

For a simple and inexpensive toilet bowl cleaner, mix ½ cup (120 ml) of distilled white vinegar and ¼ cup (55 g) of baking soda (you can double these ingredients if needed). Pour the vinegar into the bowl and let it soak for 15 minutes. Then add the baking soda and scrub the fizzy solution with your toilet brush. Neither ingredient will harm the plumbing, or more important, your septic system.

INSECT CLEANUP AND REPELLENT

Create a simple glue trap by mixing three parts corn syrup and one part water. Heat this mixture and drip it on to strips of cardboard. Set these glue traps in tight closets, behind couches, or by bed headboards and sideboards where there is contact with a wall surface.

Or, try this homemade solution from the Department of Entomology at Clemson University, South Carolina. Puree 2½ whole lemons (including rind) in a blender. Combine this concentrate with 1 gallon (3.8 L) of water and apply it with a spray bottle. Use around pet food bowls, counter-top canisters, and door thresholds. For spider control, knock down the webbing first, then apply the solution. They will not rebuild, even with little vigilance on your part. The spray smells great and does not stain or leave a film, but always test on a sample area first.

The Power of Baking Soda

Cost savings

Between $0.15 and $4.00 per use, depending on the application

Benefits

Safe, effective, and low-cost household cleaning in many areas

Some of the simplest household ingredients make the most versatile cleaners, and baking soda is at the top of the list. Baking soda, or sodium bicarbonate, is a mildly alkaline mineral made from soda ash. We use it in cooking, as a digestive aid, and as a deodorizer. But its applications extend beyond that to mildew chaser and drain cleaner, to name a few.

Baking soda is inexpensive, nontoxic, and compatible with many ingredients and materials we come across in the home every day. As a deodorizer, it neutralizes the odors of sour milk, nausea, and spoiled fruit; it also absorbs odors in the refrigerator, laundry, and in sink drains and dishwashers. When purchased in bulk, it's relatively cheap at about 60 cents per pound. Following are a list of cleaning jobs you can manage with the help of baking powder.

GENERAL-PURPOSE CLEANER

Mix baking soda and water to clean and deodorize countertops, appliances, utensils, pet toys, and most kitchen surfaces. The mixture is soft and only mildly abrasive, so risk to aluminum, chrome, silver, or porcelain surfaces is minimal. It is especially safe on tile and fiberglass. Mix 1/4 cup (55 g) of baking soda in 1 quart (946 ml) of warm water and use a cloth or nonabrasive sponge to clean items and surfaces.

DEODORIZER

Deodorize a kitty litter box by sprinkling 1/4 cup (55 g) or more of baking soda into the box. Freshen a refrigerator by placing an open box of baking soda at the rear shelf where circulation occurs. (One box will last several months.) Don't fall for the prepackaged ventilated box made for the refrigerator; it's cheaper to just replenish the baking soda from your own supply. It also works as a daily maintenance air freshener. Keep a small bowl or tin filled with baking soda on a shelf or some other inconspicuous spot.

LAUNDRY DEODORIZER

Remove strong odors from clothing by adding 1/2 cup (110 g) of baking soda to your regular wash load. (Avoid soaking silks and wool in baking soda since they are sensitive to alkaline.) The trick is, 6 to 10 minutes during a wash cycle isn't enough time for baking soda to work effectively: you need a longer soak time. So agitate the soap and baking soda mixture briefly, then allow it to rest through a presoak cycle. Or, start the wash cycle, shut off the machine, and allow the clothes to soak for most of a day. Reagitate every few hours to mix the solution. Complete by allowing the washer to finish its cycle.

DRAIN SANITIZER AND DEODORIZER

A weekly regimen of pouring 3/4 cup (165 g) of baking soda into the kitchen sink drain will neutralize odors and keep the plumbing free of clogs and buildup. Smaller bathroom sinks may only need 1/2 cup (110 g) while larger sinks can benefit from a full cup (220 g). Do not flush the powder away all at once; allow it time to work by trickling a bit of water down the drain. Kill two birds with one stone by adding baking soda to the drain, then sprinkling a bit onto the sink surface; scrub the sink with a wet cloth or sponge while allowing the used water to trickle down the drain. After 5 or 10 minutes of soaking, you can rinse the sink and drain with fresh, warm water.

CUTTING BOARD STAIN REMOVER

A dusting of baking soda onto a wet cutting board will help remove stains from acid foods such as tomatoes and coffee, and it will deodorize and even help to sanitize the surface. Allow the powder to soak in for about 30 minutes or so, then rinse the board in hot water before drying.

CARPET DEODORIZER

Scented commercial carpet and rug deodorizers are expensive and unhealthy. Instead, use baking soda. Sprinkle a 1-pound (455 g) box (more if you need it for a large carpet) evenly over the rug surface and allow the powder to sit for at least 12 hours. You can add a teaspoon (5 g) of potpourri to the powder if you want a mild scent. If the baking soda is heavier in some places than others, sweep it with a clean broom for even coverage. Vacuum the powder up; it's best to start off with a fresh filter bag.

OVEN CLEANER

Dust the lower pan of the oven with a generous layer of baking soda and moisten the powder with a water spray. If the baking soda begins to dry, moisten it again as needed. Allow the paste to work for 10 or 12 hours, then remove it with an old spatula or a coarse rag. Rinse repeatedly until the powder residue is gone.

FOOT SOAK

Mix warm water and 1/4 cup (55 g) of baking soda in a foot tub for a relaxing and deodorizing soak for sweaty, tired feed. Add a few tablespoons of Epsom salts for deep relief. Soak until the water cools.

Make Your Own Toilet Bowl Cleaners

Cost savings

Can be substantial, as commercial cleaners are expensive—up to $0.82 per treatment, or $3.28 per month

Benefits

Cleaning and disinfecting without the use of harsh and toxic chemicals

Cleaning the toilet is a grungy job, and the commercial cleaners on the market can make the task downright toxic. The good news is that you can actually make an effective toilet bowl cleaner using household ingredients that are a lot less forbidding than the cleaners you buy at the store; in fact, many of them are completely edible.

Commercial toilet bowl cleaners sold at the grocery store almost always contain substances toxic to our health and harmful to the environment, regardless of what the marketing pitch says. Avoid using bowl cleaners that contain ammonia, chlorine bleach, or hydrochloric acid. (And, don't expect the label to tell you that; manufacturers are not required to list ingredients.) The tipoff is when the back label print includes language such as "poison," "corrosive," "danger," and the like.

Instead, go to your kitchen cupboard for distilled white vinegar, baking soda, and borax. These ingredients, combined with other household items, make bowl cleaners that work well and do no harm. Plus, they cost a lot less than the chemicals you'd buy at the market.

Vinegar Bowl Scrub

- ⅓ cup (73.6 g) baking soda
- ½ cup (120 ml) distilled white vinegar
- Several drops Dr. Bronner's or similar scented castile soap

Dust the inside of the toilet bowl with baking soda so it covers as much of the surface as possible. Pour the vinegar into the bowl and allow it to soak for 10 minutes. Add several drops of a scented castile soap (eucalyptus, lavender, peppermint), and scrub with toilet brush. The mix can be flushed or left in the bowl until the next use.

Borax Bowl Scrub

- ¾ cup to 1 cup (177 to 225 g) borax powder
- Several drops Dr. Bronner's or similar scented castile soap (optional)

Pour the borax into the bowl and allow it to soak for several hours. You may add a drop or two of scented castile soap for deodorizing, if desired. Scrub the bowl with a toilet brush. Borax is not only a mild disinfectant but will also remove iron stains and scale from the porcelain.

Pumice Stone Scrub

- Pumice stone
- ⅓ cup (73.6 g) baking soda

A pumice stone is a block of light, porous, volcanic material, which when ground produces a soft but abrasive powder, often used in cleansers. Long-standing stains in the bowl can be difficult to remove with a brush alone, so a physical scrubbing is sometimes needed. Pumice stones come mounted on a handle or can be used directly with heavy-duty rubber gloves. Pour the baking soda into the bowl, scrub with a toilet brush, and use the stone to rub away tough stains. Do not scrub too aggressively, as this can scratch the porcelain finish. Spots under the rim and at the water line are good targets for pumice; borax can be substituted for baking soda if desired.

Make a Spa from Your Pantry

Cost savings

Between $7 and $16, depending on treatment

Benefits

Low-cost therapies from common household items; softer, smoother skin; and bouncier hair

Although smearing food on your face might *sound* ridiculous, the most expensive spas and beauty products boast that they use lemons, olive oil, honey, and milk. Instead of paying top dollar for a small amount of those foods diluted with fillers, why not use those ingredients at home? The following suggestions are favorites of Sara Trachtenberg, one of *BackHome*'s contributors.

The recipes described here have been tested. Part of the fun is experimenting with ingredients and coming up with something you really like.

MORE THAN SKIN DEEP

For washing, moisturizing, and replenishing the face, some simple kitchen ingredient can replace expensive products sold at beauty counters.

Powdered milk: Powdered milk is a gentle face wash that provides mild exfoliation. Put little powder in your palm, and add water to make a paste. Gently rub your face, then rinse well. I keep an envelope of it next to my sink and wash before I shower. This way all of it, even the drips, will be rinsed off, and I won't smell like sour milk.

Vegetable shortening: This is the cheapest body moisturizer you can use.

Oatmeal: For an effective facial mask or scrub, grind oatmeal in the blender to form oat flour. You now have colloidal oatmeal. Either by itself or mixed with wheat germ, oat flour makes a perfect cleanser and scrub in one. To use as a mask, mix it with milk or water to make a paste and smooth it on your face. Let the paste rest for 10 to 15 minutes and rinse it off. The mask will drip and get a little messy, so drape a towel around your neck. Another option: add a little honey to the mask, which has nourishing benefits as well as the stickiness to keep that oatmeal on your face.

Citrus peel: Never throw away a citrus peel. Remove the colored part with a paring knife and dry this zest, first in the microwave a few minutes, then on the windowsill for a couple days. Grind it in a blender or coffee grinder. The result is a free and aromatic scrub that's ideal for the face, especially oily skin.

Lemon juice: For a pore-shrinkin toner, mix lemon juice with water and splash it on your face. (Don't rinse off.) Or, try mixing lemon juice with sugar to make a paste that is full of alpha hydroxy acids and makes pores appear smaller. Rub it in gently, focusing on your chin and your nose, and rinse.

Baking soda: Soften your bathwater with a cup (220 g) of baking soda. Make a paste to use as a facial mask for oily skin. (Wear the mask for 10 minutes.)

Cucumbers: Cucumbers are great for the skin. Slide a couple of thin slices over your skin to minimize pores. Place a slice over each eye to reduce puffiness and redness. It may be something you've seen in old movies, but it does work and uses the same principle as expensive eye gels that contain cucumber and lots of carriers.

Honey: Honey is an antioxidant and antibacterial agent. Wear it as a facial mask for 10 minutes or dab it on pimples for an effective emergency fix; it dries them out and heals them. Leave on acne for 10 minutes, then rinse. Add about a cup to your bathwater for silky smooth skin.

TUB AND SHOWER TREATMENTS

Nothing exfoliates your body like plain cornmeal. It is messy, so use a hair trap in your tub and empty it while you rinse if the cornmeal prevents the water from draining.

To make another good scrub, combine equal parts sugar and olive oil, perhaps with lemon juice. This is what spas use and it feels like a great back scratch.

One of my favorite masks is yogurt mixed with an equal amount of brewer's yeast. Mix up a tiny amount, about a tablespoon (15 g), for the face and about half a cup (115 g) for your entire body. Both are supernourishing skin foods that prevent breakouts and blemishes. When wearing a body mask such as this, I have found it best to turn on some music, slather on the mask, and stand in the tub with the timer set for 10 minutes, then shower off. The skin on your body can benefit from a nourishing mask as much as your face.

YOUR SHINING GLORY

Lots of home beauty treatments for hair, such as banana and avocado, are impossible to get out! I had to pick dried banana out of my hair all day, which was embarrassing and not worth the trouble. Other treatments such as yogurt, mayonnaise, and olive oil nourish your hair without as much mess.

One of my favorite hair treatments is two egg yolks beaten with an equal amount of rum. Slather enough of a treatment on your hair to cover it, and cover with a shower cap. (Or you could bunch up a plastic bag and secure it with a rubber band.) Leave on for 20 or 30 minutes, and wash out with cool water. (Wash eggs out of your hair in cool water, or they will scramble.)

Try mixing an equal amount of honey with your shampoo for moisturizing and body. Or rub a tablespoon of honey on your scalp and wet hair before shampooing for an immediate cleansing. You will feel the difference as it dissolves residue from commercial products you've used.

Be penny wise and buy the least expensive shampoo you can find. After years of experimentation, I believe that all the ones on the market work pretty much the same. Simple add some pantry items to your dollar-store shampoo to make a superior product. Apple cider vinegar or baking soda are good choices for getting rid of residue; add one part to two parts shampoo. Or you might add a small amount of olive or other oil for nourishment. To add bounce to my hair, I mix a small amount of flat beer into my shampoo.

Make Home-Grown Skin Care

Cost savings

Depending on the affliction and the medicine used to heal it, savings can run from about $6 to more than $35

Benefits

Making medicinal remedies at almost no cost from household ingredients

Natural skin treatments are desirable for a number of reasons, the least of which focus on comfort and appearance. Here, *BackHome* contributor Dr. Charles Dickson addresses this aspect of commonsense, homegrown health care.

The skin is the largest organ of the body, and the average adult has 17 square feet of it. Taking care of our skin is vital because the skin is the first line of defense against any invasion of foreign matter that can be harmful to the human system. The skin is often considered to be a third kidney, because we excrete toxins through the skin pores just as we excrete through the kidneys. We also ingest many chemicals and toxins through the skin, so it is important to pay attention to the way we cleanse, soothe, and heal our skin. In addition, the skin plays an important role in regulating body temperature.

But skin also has issues: acne, warts, poison plant rashes, insect bites, and stings. In treating these conditions with home remedies, we need to know which remedies work, understand why they work, and learn how to make them properly.

Why do skin problems surface? Some issues are due to internal factors, such as poor diet. By following a simple, natural diet and avoiding processed foods, our immune system will stay healthy and better manage bacteria and viruses. There are external issues, like accidentally walking through a patch of poison ivy.

Either way, there are effective home remedies you can make in your own kitchen "pharmacy" to clear up skin problems.

ACNE

Acne can be a problem at any age. Try one or more (not at the same time!) of the following natural aids.

- Combine 4 ounces (115 g) of grated horseradish with a pint (475 ml) of 90 proof alcohol. Add a pinch of nutmeg and bitter orange peel, and apply to the areas affected by the acne.
- Mix the juice extracted from two garlic cloves with an equal amount of vinegar. Apply several times a day; the condition should clear up in a couple of weeks.
- Place ½ cup (80 g) of chopped onions in 1 cup (235 ml) of olive oil. Allow the mixture to steep for 1 week, then strain it through cheesecloth into a bottle. Apply to the acne, and leave it on for 15 minutes before wiping it off.
- Steep two papaya mint tea bags in ½ cup (120 ml) of boiling water. Let the tea bags steep until the tea is very strong. While the water is still hot, soak a washcloth in it, and then apply the washcloth to affected areas as a hot compress.

WARTS

Warts on the skin are believed to be caused by a virus, and removing these unsightly spots can sometimes be very challenging. Here are some solutions.

- Apply a wet tea bag to the wart for 15 minutes a day. Many report that the wart is gone in a week to ten days.
- Grate some carrot and mix with a little olive oil. Place this mixture on the wart for a half hour twice each day.
- Dab on some lemon juice, followed by some raw chopped onion. Let this stay on the wart for 15 minutes each day.
- Crush a fresh fig and place it on the wart for a half hour twice each day.

POISON IVY, OAK, AND SUMAC

The irritating substances exuded from poison ivy, oak, and sumac all cause skin eruption, particularly for those whose body chemistry is such that they are highly allergic to the urushiol and other related chemical compounds in these plants. The following homemade remedies deal with the itching and discomfort.

- Add 1/2 cup (120 ml) of apple cider vinegar to 1/2 cup (120 ml) of rubbing alcohol. Mix well and apply to the affected areas twice each day.
- Chop four garlic cloves and boil them in 1 cup (235 ml) of water. Cool and apply this mixture with a clean cloth to the affected skin.
- Rub the inside of a banana peel directly on the rash once each hour for two days.

INSECT BITES AND STINGS

Different people react in different ways to bites and stings from insects. Most stings contain a chemical called formic acid that can cause anything from minor irritation to an acute medical emergency. If bites and stings cause only minor discomfort, the following steps can be taken to alleviate the pain.

- Place a slice of raw onion or raw potato on the location of the sting to draw out the poison. Leave the vegetable there for a half hour and follow this with an ice cube to reduce swelling.
- An equal mixture of vinegar and lemon juice (which respectively contain acetic and citric acids) will go a long way toward relieving itching and swelling.
- Wet a clump of tobacco and put it on the sting.
- Put a dab of honey on the sting.

PSORIASIS

This chronic inflammatory disease of the skin is usually characterized by silvery scaling patches. Dealing with this problem can sometimes be discouraging, but not hopeless. Try these remedies.

- Rub garlic oil on the area several times each day.
- Squeeze the juice from an aloe vera leaf and rub it on the affected skin.

Make Homemade Baby Care Products

Cost savings

Between $11 and $19 compared to store-bought products

Benefits

Enjoy a safer, cheaper alternative to store-bought baby care products

One way to ensure the purity and quality of the products you use on your baby or child is to make them yourself. Here, *BackHome* contributor Maggie Julseth Howe presents these thrifty homemade baby care items.

Babies and children have sensitive skin, newly exposed to the harshness of life, so it's important to consider what products we use on them. Babies and children also love to put their hands (and feet!) in their mouth, so moisturizers, powders, or lotions could be ingested. The lotions, soaps, creams, and powders that we use on children should be as natural and healthy as possible.

Babies and children often have dry skin, and most baby lotions, salves, and oils contain high percentages of mineral oil or other petroleum products. Mineral oil does not absorb well into the skin, and in fact clogs pores and makes it difficult for skin to eliminate toxins. It's preferable to use natural vegetable oils as the base ingredient in lotions, salves, and moisturizing oils.

Many baby products also contain colorants, fillers, and fragrances that do little except irritate sensitive skin. Synthetic fragrances can exacerbate baby eczema, acne, and diaper rash. Check out any store's baby care aisle and you'll see that all these lotions and potions are also quite expensive.

One way to ensure the purity, freshness, and quality of the products you use on your children is to make them yourself. The recipes that follow are simple to make and use common yet effective all-natural ingredients to inexpensively create exquisite formulas that will pamper and soothe your baby.

Moisturizing Milk Bath

Milk is a soothing, moisturizing, nongreasy bath additive that gently cleanses your child's skin. Chamomile and lavender essential oils are calming. A warm lavender aromatherapy bath does wonders for a fussy baby or toddler! (Please use essential oils only on babies 6 months or older and keep the raw oils far out of your little one's reach to avoid the risk of ingestion.)

- 1 cup (52 g) dried milk
- ½ cup (64 g) cornstarch
- 2 to 3 drops lavender or chamomile essential oils (optional)

In a small bowl, combine all ingredients and stir. To use, sprinkle a small amount in a warm bath. This makes a wonderful gift, especially when presented in a pretty recycled shaker jar or bottle.

Baby's Bum Powder

This talc-free powder utilizes the power of nature's finest herbs to gently soothe and heal irritated skin. Arrowroot powder (found at health food stores) gently helps soothe and dry damp areas, while powdered lavender and chamomile are healing and help prevent bacterial growth.

- 1 cup (120 g) arrowroot powder
- 1 tablespoon (7.5 g) dried ground chamomile (see note)
- 1 tablespoon (7.5 g) dried ground lavender (see note)

Combine the arrowroot, chamomile, and lavender in a plastic sandwich bag, seal shut, and knead the bag with your hands. Pour the powder into a recycled shaker jar or powder tin. Use as needed on the diaper area.

Note: Grind chamomile and lavender into a fine powder using a clean coffee grinder, food processor, or blender.

Better Baby Oil

Vegetable oils are more nourishing, soothing, and moisturizing than mineral oils. Vitamin E is a natural antioxidant and is soothing to dry, chapped, or irritated skin. This makes an excellent moisturizer and also works well to loosen and heal cradle cap or eczema. The many benefits of infant massage are well documented, and this oil also makes an especially soothing natural massage oil.

- 1 cup (235 ml) grapeseed, almond, sunflower, or olive oil
- 2 or 3 capsules vitamin E

Place the base oil in recycled squeeze bottle. Cut open or pierce the vitamin E capsules and squeeze the vitamin E oil into the base oil. Stir or shake the oils until combined. Use as needed in the bath, or as a lotion or massage oil.

Herbal Baby Balm

The healing powers of the herbs in this salve are safe and gentle yet powerfully effective. They are transferred to the oils, and then beeswax is added to make a creamy balm. Beeswax naturally helps seal moisture into the skin while still allowing it to breathe. You can use this salve to help relieve diaper rash, heal cuts or scrapes, or act as a moisturizer or lotion for dry skin.

- 2 cups (475 ml) olive, sunflower, grapeseed, or almond oil
- ⅓ cup (43 g) dried comfrey
- ⅓ cup (43 g) dried chamomile
- 2 ounces (approximately 2 tablespoons or 57 g) beeswax
- 6 capsules vitamin E

Combine the oil, comfrey, and chamomile in a heavy-bottomed saucepan. Place on the lowest possible heat; the oil should be warm but not hot enough to "fry" the herbs. Let the oils and herbs infuse for at least 3 hours. Let the oil cool, then strain out the herbs. Gently rewarm the oil, cut or pierce the vitamin E capsules and squeeze into the mixture, and add the beeswax. Stir until the beeswax is completely melted, then remove from the heat. Pour the salve into clean jars or tins.

There also are excellent commercial natural baby products. Unscented, handmade soaps (especially castile or goat's milk soaps) can often be found at farmers' markets or natural foods stores. Read labels and look for all-natural products free from petroleum products, mineral oils, synthetic colors, and synthetic fragrances. Both your child—and you—will be happier and healthier.

Save a Flush

Cost savings

Averages about $9.26 per month for a municipal system

Benefits

Water reuse saves money and our most valuable resource

The average family uses more water for their toilets than for any other household appliance. *BackHome* contributor Tim Burdick has come up with a solution that not only saves potable water, but reduces the water bill as well.

In our house, we have a secret in our bathroom. No one in our extended family or network of friends knows about this secret. We keep it safely hidden from our guests when they come to stay and have even avoided discovery by our babysitter. Despite this cloak-and-dagger approach, this mystery saves a few dollars on our water bill and helps us do our part to avoid wasting water. Our secret? A one-gallon (3.8 L) plastic water pitcher we bought on sale for $2.

We call it The World's Most Inexpensive Gray Water Recycling System.

Here's why the system is worth considering: The average American family of four uses about 700 gallons (2,650 liters) of water a week in their toilet use alone. If 50 families implemented a similar style of gray water recycling system just one day a week, enough water would be saved over the span of a year to fill an Olympic-size swimming pool.

The fact is, in most homes, even low-flush toilets are the single largest consumer of water, more so than anything else (shower, dishwasher, washing machine, sinks, etc). Anything we could do to satisfy this thirsty monster would help save money and resources.

Our family's system had to be flexible and uncomplicated. When we have guests over and there is a higher volume of flushing, we didn't want to run out of water. We didn't want our guest feeling embarrassed as they stood there wondering why their flush didn't go down. We needed a system that could handle sudden changes in usage and could easily be turned on or off.

Now you can buy commercial gray water flushing systems to solve these problems, but they range from several hundred to several thousand dollars, not to mention installation fees. In lieu of such a costly venture, the solution for our family turned out to be a $2 water pitcher and a shower drain plug. Now, whoever takes the last shower each day simply puts the drain plug in and lets the tub fill up with water. For the rest of the day we have an instant source of nearby gray water that gets bucketed over to the toilet for each flush.

You simply dump the water directly into the bowl at a quick rate and it flushes just like normal. In our home, the toilet is less than 2 feet (61 cm) from the shower, so it takes minimal action to reach down for a scoop of water and repour it into the toilet. Usually we only need about 1 gallon (3.8 L) of water for each flush, which is less than the standard 1.6 gallons (about 6 L) used in low-flush toilets. At the end of the day, any remaining water still in the tub is drained away, and we ensure that our last toilet flush of the night is a fresh water flush.

By doing it this way, we have none of the problems associated with storage (the water is used up or drained out within a 12-to 14-hour period), there is no electricity or pumping involved, and the system is flexible. On days when we have company, we simply don't fill up the tub, and everyone uses the normal method of flushing instead. As a side benefit, we also return free heat back into the house as the shower water cools.

Just to be safe, there are some additional precautionary measures we take as a family to ensure the system works well. Mainly, the shower and toilet get cleaned and disinfected twice a week now, instead of the traditional once.

As mentioned before, our last toilet flush of each night is a fresh-water flush to eliminate any storage concerns (gray water sitting in a toilet bowl for long periods of time becomes rancid). We also run the water pitcher through the dishwasher every so often to make sure it is sterilized. Additionally, if a large amount of biowaste is put into the toilet at one time, this method may not send it all down on the first try. Sometimes we do a second pitcher of water or flush normally.

Finally, for some people, the idea of standing in used shower water may not seem hygienic, but it differs very little from taking a bath, and one could still easily rinse off their feet with fresh water when stepping out of the shower. Besides, you get the joy of a guilt-free long shower.

In the end, for us, this solution turned out to be the world's most inexpensive (and highly effective) gray water recycling system. Fortunately, it can easily be replicated, and probably should be. The beauty of this idea is that it can be implemented today by just about anyone for practically no cost. Don't worry if you are a little shy about sharing your bathroom secret with others. If company comes over for a surprise visit, pull the drain plug, stash the water pitcher in the cabinet, and everything is back to normal.

Become Septic System Savvy

Cost savings

Up to $6,000 for a replacement septic system

Benefits

Maintaining a healthy septic system extends its life significantly and prevents environmental damage

Most people have no idea how a septic system works. In fact, some people don't even know they have a system, which is why failing sanitary systems are a serious environmental and health problem. Poorly functioning septic systems are the leading source of waterborne disease outbreaks. And septic system replacement can be expensive, with costs running from $3,000 to $6,000 and more.

Briefly, here's how a typical system works: Water that comes into the house is used for cooking, washing, showering, and flushing. Both the drained and flushed water goes through the household waste pipes and out a large line to the septic tank, buried about 10 or 20 feet (3 to 6 m) from the house. The container holds between 1,000 and 3,000 gallons (3.7 and 11.4 kl) and is designed to separate solids from liquids while encouraging the bacterial decomposition of the solid material.

The liquids, mostly watery effluent, flow out of the tank into a drain field, which is a series of perforated pipes buried in a bed of gravel and covered with earth. Microorganisms in the surrounding soil eliminate any pathogens in the water that were not destroyed by the bacterial digestion process in the tank.

When a well-designed septic system fails, it's usually the soil, not the tank, that is at fault because it won't allow liquid to pass through. In most cases, the soil fails when it becomes plugged with solids. Following are some simple points to be aware of to ensure that your actions are part of the solution and not part of the problem.

WATCH YOUR WATER USE. Doing a large number of laundry loads in a short period of time can damage your system. Typically, solid materials settle in the tank while effluent flows out into the soil to the filtered and cleansed. If you put more water into the septic system than it is built to handle, the water will flood the system and may flush solids out of the tank into the drain field, where they will clog the soil. Spread out your water use. Do one or two loads of laundry per day, rather than a dozen loads on a Saturday morning.

Mechanical water softeners can also damage your system by putting too much water through the septic system. Upgrade your softener with a newer efficient model, or install a separate mini-septic system for your softener.

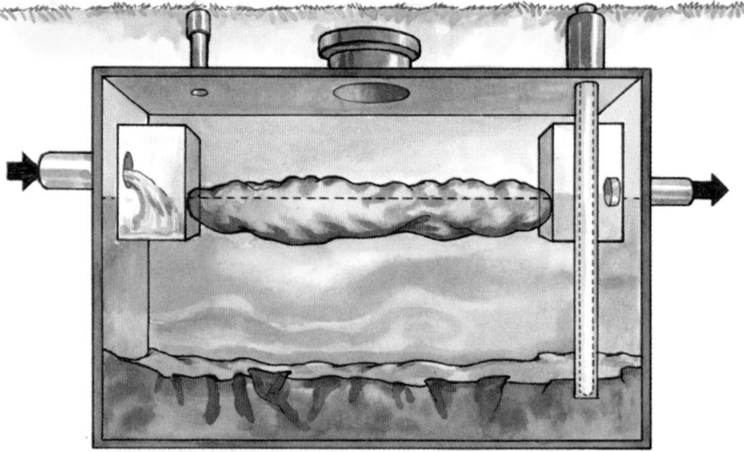

Prevent solids from leaving the tank. You should get your tank pumped at regularly scheduled intervals to prevent excessive accumulation of solids in the tank. This means every couple of years, and in some cases more frequently. Tanks should be pumped and inspected through the manhole cover, not the inspection pipe.

An effluent filter (about $70) can be installed in the exit baffle of the tank. It prevents the larger solids from getting out to the drain field. These filters are cleaned every few years when you have your tank pumped.

Limit use of household cleaners. Excessive use of chlorine bleach, detergents, and lye-based cleaners can contribute to septic system failure. If you do more than five loads of laundry a week using bleach, problems may arise. Avoid powdered detergents, because they contain nonbiodegradable fillers that can plug up your system. Also, be careful with harsh automatic toilet bowl cleaners.

Be careful what you flush. There are other things you should never flush or put down the drain. These include motor oil, paint thinner, oven-cleaning agents, fats, pesticides, kitty litter, coffee grounds, and disposable diapers. Root inhibitors and similar products containing copper sulfate meant to destroy invasive growth also kill the beneficial organisms that break down waste in the system. And even septic tank additives containing enzymes are not particularly beneficial. They tend to break down standing sludge into smaller particles that are then flushed out into the drain field, where they can clog the system.

Catch your washing lint. Lint generated in the washing cycle clogs the soil in drain fields. The typical family washing machine produces enough lint every year to carpet a floor. Lint screens trap only a fraction of the lint fibers. These minute particles stay in suspension and are flushed out to the drain field, where they plug up the pores of the soil bed. Lint can be prevented from entering the septic system through the use of a reusable in-line filter, which attaches to your washing machine discharge hose.

Don't use a garbage disposal. Food and kitchen waste can usually be disposed of by outdoor composting or through your garbage service. The convenience of an in-sink garbage disposal can increase the volume of solids in the system by as much as 50 percent. Undigested food scraps build up solids and must be pumped from the tank more frequently.

Here are a few things you can do to maintain the health of your septic system and the drain field in particular.

- Discharge only biodegradable wastes into system.
- Divert surface runoff water from roofs, patios, driveways, and other "hardscape" areas away from your drain field.
- Keep your septic tank cover accessible for routine maintenance.
- Have your septic tank pumped regularly. It should also be inspected for leaks and cracks.
- Don't dig in your drain field or build anything over it.
- Don't drive over your drain field or compact the soil in any way.
- Don't plant trees or shrubbery in the absorption field area, because the roots can get into the lines and plug them. Grass is the only thing that should be planted on or near a drain field.
- Don't cover the absorption field with hard surfaces such as concrete or asphalt.

CHAPTER 4

BEDROOMS AND GUEST ROOMS

Bedrooms often serve as a second home—at least for your guests. So it makes sense to keep them clean, comfortable, and welcoming. This chapter provides ideas for freshening up the space with homemade aromatics and some strategies for organizing the room and saving energy.

Bring Spring Indoors with Natural Aromatics

Cost savings

Most of these recipes cost pennies on the dollar compared to store-bought items

Benefits

You choose what scents you prefer, and the concoctions are inexpensive

There's something satisfying about welcoming the season with a clean, fresh home. But there's no need to stock up on expensive commercial products when safer, more effective mixtures can be made at home. Here, Sandra Noelle Smith offers some natural recipes for freshening up the home.

You'll save money if you're able to harvest the herbs used in these recipes right from your own garden. If you don't grow the herbs and the recipe sounds good to you, make a note to add these plants to this summer's garden.

Scented Furniture Polish

This is great for unvarnished furniture.

- 3 ounces (85 g) unrefined beeswax, grated coarsely
- ¾ cup (175 ml) natural turpentine
- A few drops of essential oil of your choice (try lemon, rosemary, lavender, sandalwood, or cedarwood)

Place the beeswax in a clean jar with a screw-top lid. Pour the turpentine over the wax. Close the jar, and let stand for approximately 1 week, stirring occasionally, until the mixture becomes smooth. Add the essential oil sparingly to scent, and blend well. To use, rub the polish into furniture with a soft rag. Wait a few minutes, wipe off, and buff to a shine. If the polish hardens over time, set the jar in a bowl of hot water.

Furniture Oils

There are several variations on the oil-lemon theme; here are two. Shake well before use.

- ¾ cup (177 ml) vegetable or olive oil
- ¼ cup (59 ml) lemon juice, or 1 teaspoon (5 ml) lemon oil
- 2 cups (500 ml) mineral oil
- A few drops essential oil (choose a complementary fragrance such as orange or lemon)

Combine the ingredients in a small glass jar and shake well. Rub into furniture with a soft rag.

Note: As with most polishing products, these may not be safe for all finishes. Test first on a small, hidden area.

Herbal Infusions for Linens

- 2 ounces (57 g) dried, or 4 ounces (114 g) fresh, of any one of the following herbs: sweet marjoram, lemon balm, lemon verbena, lavender, or mint

Place in a heatproof glass or plastic container (do not use metal). Pour 1 cup (235 ml) of boiling water over the herb. Cover, and let stand for about 3 hours. Strain off the liquid into a glass jar. Sprinkle a few drops of the infusion over bedsheets, pillowcases, etc., or pour the infusion into a spray bottle to give your linens a fine mist.

Note: Dark infusions may stain light fabrics. As an alternative to sprinkling directly on the fabric, dampen a handkerchief with the infusion and allow to dry. Fold and tuck into your pillowcase.

Door Sachets

These are simple enough to make for every room in the house. Hung from a doorknob, the sachet perfumes the room every time the door is opened or closed.

- approximately 14 x 3 inch (35.6 x 7.6 cm) piece of muslin, bleached or unbleached
- Cotton thread for sewing
- Potpourri for filling (see page 101)
- Large cotton ball
- 3 to 4 drops essential oil
- Rubber band
- 20-inch (50.8 cm) length ribbon, twine, or raffia

Fold the muslin in half lengthwise, wrong-side out. Sew the long edges with a ¼-inch (6 mm) seam, creating a pouch about 7 inches (17.8 cm) long and 2½ inches (6.4 cm) wide. Turn down the fabric at the top to form a 2-inch (5.1 cm) fold, and iron in place. Turn the pouch right side out, using the eraser end of a pencil to push the corners out. Fill with potpourri to about 1 inch (2.5 cm) from the top. Drop the oil onto a cotton ball and bury in the center of filling. Close the neck of sachet with a rubber band.

Use ribbon, twine, or raffia to form a hanging loop, winding around the neck to conceal the rubber band. Knot securely, and trim excess ribbon. If desired, decorate with a small bunch of dried lavender or a dried orange slice tied on with raffia. You may want to make extra potpourri for "refreshing" your sachet. Just double the recipe and store the extra in a glass jar with a screw-top lid. Replace the filling when a sachet begins to lose its fragrance.

Sachet #1: Country Spice Kitchen

- ⅓ cup (75 g) white-pine wood shavings
- 2 tablespoons (28 g) cinnamon basil, dried and crumbled
- 1 tablespoon (14 g) ground cloves
- 1 tablespoon (14 g) allspice berries, cracked
- ½ teaspoon (1 g) fennel seed
- 1 cinnamon stick, broken up
- 3 drops sweet orange oil
- 1 drop clove oil

Sachet #2: Lavender-Rosemary Bed/Bath

- ⅓ cup (75 g) white-pine wood shavings
- 3 tablespoons (9.9 g) dried rosemary
- 3 tablespoons (42 g) dried lavender buds
- 3 drops lavender oil

Hint: Make an extra sachet to hang from the rearview mirror in your car.

MOTH-REPELLENT SACHETS

Time to pack away those woolens! Forget smelly mothballs; several herbs with moth-repelling properties cost less and are not a health hazard. To create moth-repelling sachets, combine a few of the herbs in presewn muslin bags, or place dried herbs in the centers of 10 x 10 inch (25.4 x 25.4 cm) fabric squares, and tie with ribbon.

Try one of the following combinations, or create your own blend. Hint: To prolong the life of sachet, protect from direct contact with rain or other moisture. Store in a sealed plastic bag or glass jar when not using.

Sachet #1

- 4 tablespoons (57 g) dried rue
- 2 tablespoons (28 g) dried mint
- 2 tablespoons (28 g) dried pennyroyal
- 2 tablespoons (28 g) dried southernwood
- 2 tablespoons (28 g) dried rosemary
- 1 tablespoon (14 g g) ground cloves

Sachet #2

- 4 tablespoons (57g) dried southernwood
- 4 tablespoons (57g) dried tansy
- 4 tablespoons (57 g) dried santolina

Other herbs to try include silver artemisia, lavender, mugwort, and wormwood.

Note: Some people are allergic to rue. You may want to wear gloves when working with this herb.

Make Space (and Save Cash) with Closet Storage Shelves

Cost savings

$950 saved by not having to hire a professional contractor

Benefits

The reuse of functional material and the satisfaction of doing a do-it-yourself project

No matter the size of a home, there never seem to be enough storage. Charlotte Anne Smith's goal was to organize the things she needed in a place where they'd be accessible yet out of the way, and she wanted to set up that space on her own to save a bit of cash.

I live in a four-bedroom, two-bath ranch that was originally home to six people but now houses only me. Still, there was a pitiful lack of storage space until I took a good look at how I was using it.

You can actually hire experts who analyze space, apply ergonomic principles, and work things out to the comfort of their clients. There is such a thing as a "professional organizer." I took matters into my own hands, and so can you.

Divide and conquer. My New Year's resolution was to transform a closet in my utility room to hold tools and paint so I could find what I need without sorting through stacks and piles. Starting with one space rather than a lofty plan to organize the whole house say, in a week, made the task manageable.

Start with a plan. First I determined what I wanted to store and what I needed to pitch. I knew I'd be using some of the excess materials to build a proper storage space, so I set aside the things that made sense to reclaim. Next, I cleaned out the closet completely. Then, using one of the cans of leftover paint, I painted the entire inside white. At this point, I had to determine exactly how I wanted the shelves spaced to hold the items I was saving and still leave room to fit a CD tower at one end and hang a cap rack at the other.

Build more storage. I cut three 12-inch (30 cm) -wide shelves from the plywood, each long enough to stretch between the two inside walls. Then, I used surplus wood to cut nine cleats to support these shelves. Six match the width of the shelf boards and three fit along the back wall. After I'd painted the lot (I used a quart [946 ml] of eggshell and a cup [235 ml] of leftover crimson to make this girl's tool shelves appropriately pink), I fastened the cleats to the wall studs along the center and at each end.

Redefine the floor. I didn't have the tools to cut ceramic tile, so a friend laid the floor for me. (You certainly don't need a tile floor. A scrap of vinyl flooring will dress up the floor, and that's an easy do-it-yourself cut-and-glue job.) With that out of the way, I set the shelves on the cleats and attached them with screws left over from a siding project.

Get organized. I was able to repurpose items I would have stored or thrown away into storage tools. For one, I used the hat rack to store extension cords and my bow saw. To make the rack fit the horizontal space between the shelves and the front wall, I turned it on its end and screwed it to the side wall, it over a wooden shim at the lower end so the pegs would tilt slightly upward and my extension cords and bow saw wouldn't slide off.

Next, I sorted all of the tools and supplies by job, so my painting things were all placed on one shelf and my pliers, hammers, saws, and so forth on another. I used the baskets in the CD tower to hold loose items so they wouldn't fall. As a final touch, I attached a wire kitchen rack designed to hold spices and other odd items to the inside of the door. Here is where I placed small plastic containers and film cans; each one contains a different size nail, bolt, screw, or fastener. The tubs that aren't transparent have either a label listing the contents or a sample taped to the outside. One unanticipated discovery was that the closet's white background made it much easier to see what I was looking for.

My goal was to organize on the cheap. By the time I was through the only items I'd bought were some cement board to go under the tile, glue and grout, and a battery-operated light for the closet. I wound up spending less than $50 for a remodel that could have cost me $1,000.

Ease Winter Woes with Heat-Saving Curtains

Cost savings

Between $25 and $100 a heating season in a home with unweatherized windows

Benefits

Fewer winter drafts and a warmer, cozier home

At one time, the cold weather solution to drafty windows and huge expanses of glass was to turn up the thermostat. That bit of convenience went out in the 1970s with the first energy crisis, beginning a move toward weather-sealed windows and insulated glass.

Nonetheless, not all homes are equipped with up-to-date windows, and even the best panes have a far lower R-value than the surrounding wall. The simplest fix is to whip up a set of insulated window shades that will retain the room's warmth and keep drafts from invading your living space and draining your purse of hard-earned resources.

Traditional Roman shades are about as basic as you can get, and they require little in the way of materials. Here's how to make shades for a 36″ x 48″ (91 x 122 cm) window.

MATERIALS

- 55 inches (140 cm) of 42-inch (107 cm) -wide shade material
- 48 inches (122 cm) of reflective thermal fleece for lining
- A roll of ¾-inch (1.9 cm) paper-backed fusible adhesive stripping
- 5 yards (4.6 m) of 1-inch (2.5 cm) -wide woven twill ring tape
- 10 yards (9 m) of 3 mm nylon cord
- A half-dozen No. 4 x 1⅝-inch (4.3 cm) screw eyes
- Several No. 6 x 1½-inch (4 cm) flathead screws
- A piece of chain for weight,
- An awning cleat
- A 36-inch (91 cm) strip of 1 x 2
- A wooden spool (or other pull)
- A staple gun
- A drill
- Sewing machine (and thread)

STEPS

Measure the window frame.
The effectiveness of the shade will depend in part on its fit within the jambs, so don't allow a gap between the curtain material and the frame. Measure the frame horizontally from inside to inside, then measure vertically from the head jamb to the sill. Write down the dimensions.

Cut the mounting board and fabric. The 1 x 2 mounting board will actually measure ¾ x 1½ inches (2 to 4 cm) to a length ¼ inch (6 mm) less than the width of your window opening. Cut a piece of shade fabric 1½ times the circumference of the board and 3 inches (7.6 cm) longer than its length (a).

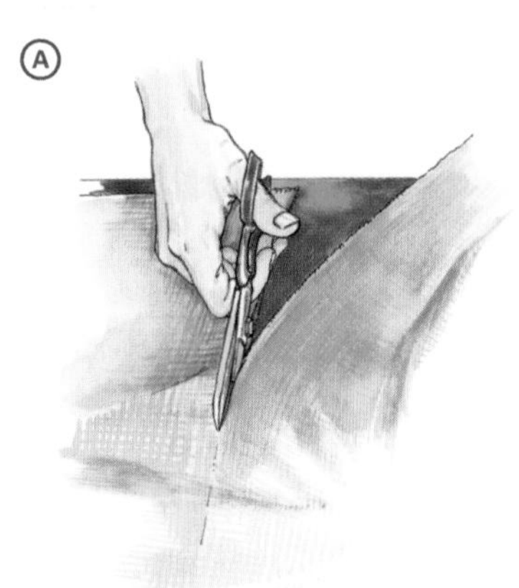

Prepare fabric board. Center the board on the fabric strip, and fold one long edge over the wood. Fasten it with ½-inch (1.3 cm) staples every 8 inches (20 cm) or so, but leave a few inches free at each end. Then, fold over the other edge of the fabric and staple it through the fabric you just fastened. Miter the fabric neatly over the ends, and staple that fabric down. Set the mounting board aside.

Cut shade to size. Lay out your remaining shade fabric on a flat surface. Beginning with the inside window dimensions taken earlier, add 2 inches (5 cm) to the width and 7 inches (18 cm) to the length to get the cut size. Use a framing square if possible to ensure your cuts will be accurate. Next, cut the reflective thermal fleece. That lining should be trimmed to the finished width (the horizontal inside measurement) and the finished length.

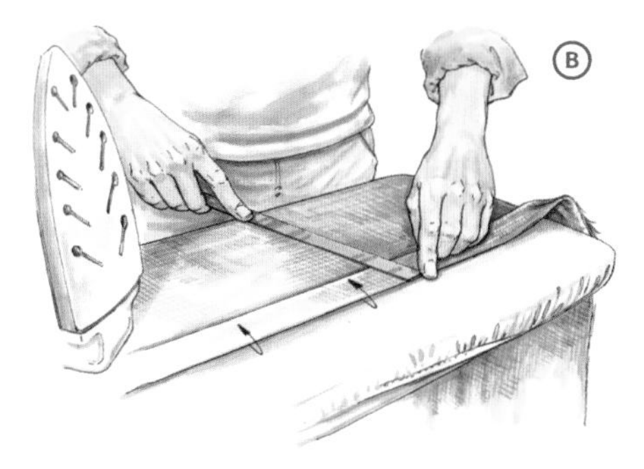

Apply adhesive. Fold 1-inch (2.5 cm) hems into the sides of the shade and press to crease with a hot iron (b). Unfold the hems and set a strip of ¾-inch (2 cm) fusible adhesive, paper side up, along the length of each side, just back from the cut edge. Trim the adhesive to length. With the paper hacking intact, fuse the adhesive to the fabric, using the tip of the iron's soleplate (c).

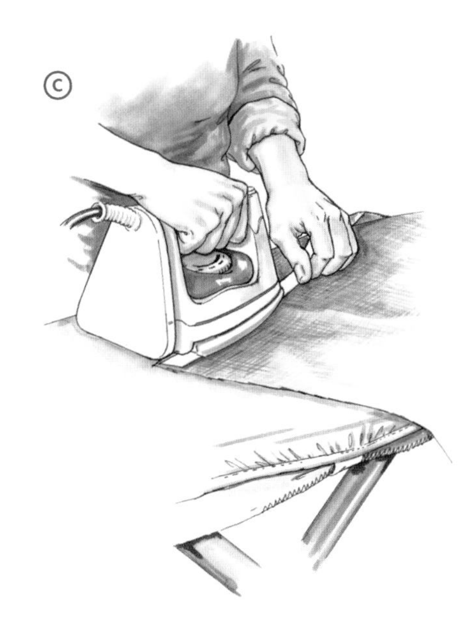

Attach thermal fleece. Align the thermal fleece lining over the fabric, wrong sides together. (The lining's reflective surface will reduce radiant heat transfer whether it is visible or not. If it faces an air space, it will be effective even if it does not face the warmest surface.) The lower edge of the lining should be 3½ inches (9 cm) above the lower edge of the shade. Turn back the shade fabric side hems over the fleece lining, pin in place (d), then remove the paper backing from the adhesive, while sealing the hems in place with the iron (e).

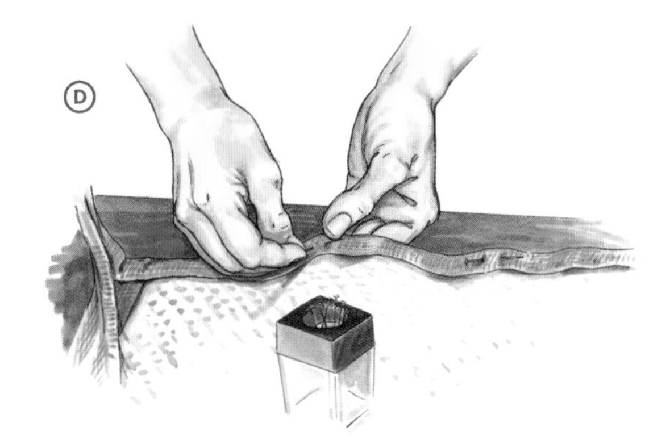

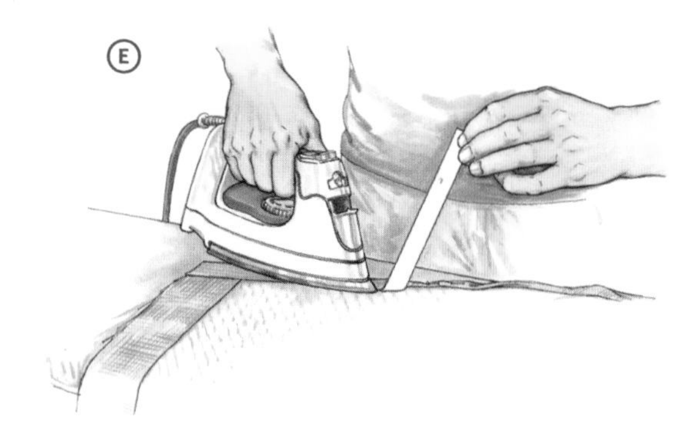

Form a hem. Fold a ½-inch (1.3 cm) hem into the lower edge of the shade fabric and crease with the iron. Then fold a 3-inch (7 cm) hem (that overlaps the lining) to form a weight pocket. Crease the edge, secure with straight pins, and stitch the ½-inch (1.3 cm) hem to the fabric.

Measure length. Measure from the bottom of the shade to the top and draw a chalk line to indicate the finished length. Cut the fabric 1½ inches (4 cm) above the line.

Cut and place woven twill ring tape. Cut four strips of 1-inch (2.5 cm) woven twill ring tape to fit between the chalk line at the top and the top of the 3-inch (7 cm) hem pocket. Heavy fabric shades or those wider than 36 inches (91 cm) may require additional ring strips, which keep the cords in position as you draw up the curtain. Space the strips evenly, 6 to 11 inches (15.2 to 28 cm) apart, with the outermost ones ¾ inch (1.9 cm) from the outside edges of the shade (f). Position the strips so the lower rings start at the top of the hem pocket, and then stitch along both sides of the tape to the top line. One row of the stitching will secure the side hem (g). (Remove any rings that are at or above the top line.)

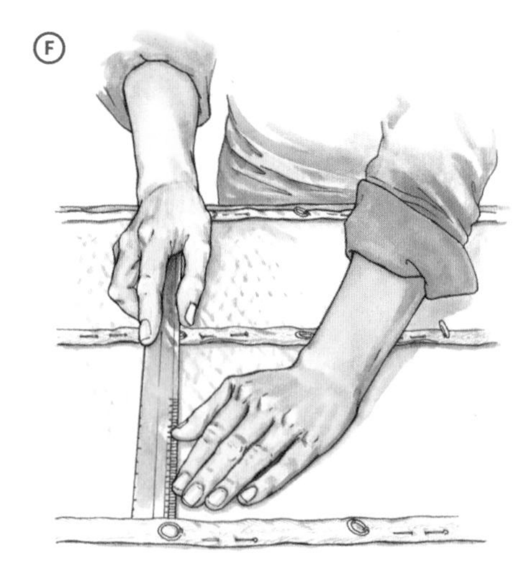

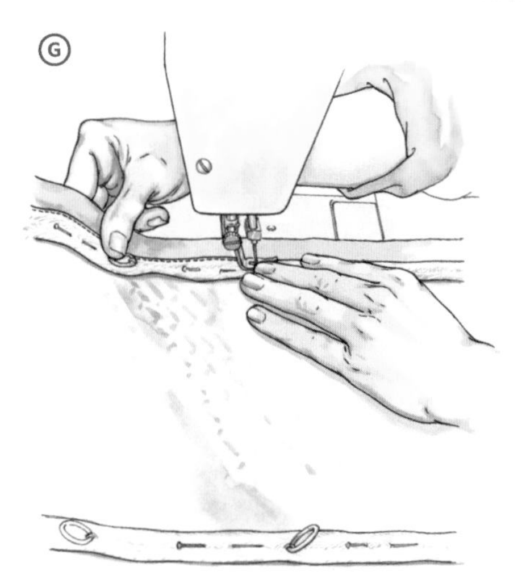

Assemble the shade. Cut a piece of swing chain or a suitable weight ½ inch (1.3 cm) shorter than the width of the finished shade. Slip it into the weight pocket and stitch the ends closed. Staple the shade to the mounting board by aligning the top line with the front edge of the board and drawing the ½ inch (1.3 cm) of extra fabric over the seam on the top of the board before stapling (h). Drill small holes into the underside of the board, lining them up with the ring strips, and install the screw eyes.

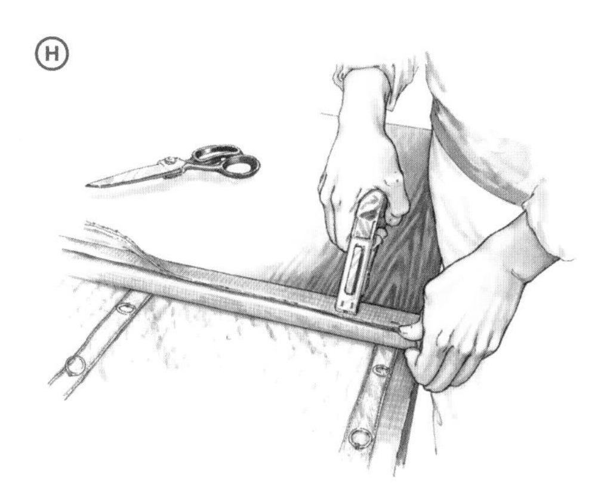

Make shade cords. Cut one length of cord for each ring strip, each 1½ times the shade length plus the distance from the strip to the drawstring side. Tie the cords to the bottom rings and string each one through its respective ring set, through the screw eyes, and over to the side (i). Pass all the strings through the knob pull and tie a temporary knot.

Install the shade. Place the mounting board against the window head jamb with the shade face-out. Starting from the screwhole side, drill a ⅛-inch (13 mm) hole through the board near each end, and one in the center, and slightly into the jamb above. Change to a smaller drill bit and continue drilling the jamb holes to a

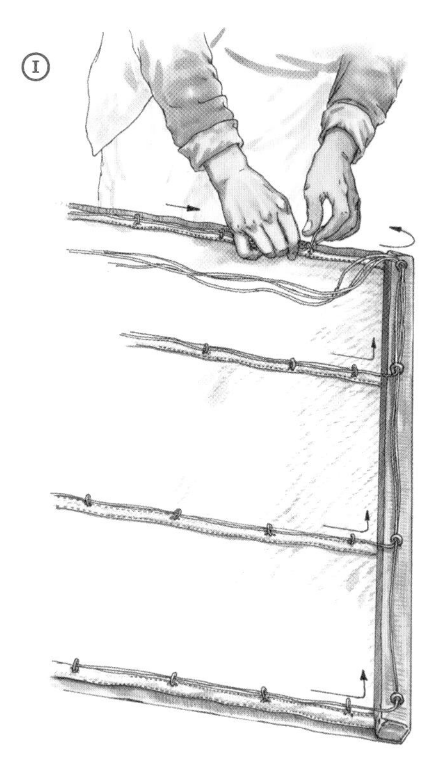

¾-inch (1.9 cm) depth. Fasten the board to the jamb with the No. 6 x 1½ inch (4 cm) screws. With the shade completely lowered, slide the knob pull up to the mounting board and knot the cords together. Trim the excess cord. Fasten the awning cleat to the window casing to keep the shade in place when it's raised.

Note: This insulated shade recipe is by no means carved in stone. The costly quilted commercial shades you see in catalogs are usually sealed to the jambs within tracks at the sides or with magnets or other fasteners. Nevertheless, if the fit along the side jambs is firm and the weight at the bottom pulls the shade down against the sill, all but the worst drafts will be stopped cold.

Save Energy with Seven Tips

Cost savings

Varies broadly, but every watt not used saves about $1 per year; savings can be greater than $40 per month in a large household.

Benefits

A cleaner, healthier environment

Energy is expensive and we use it generously in our homes. In winter, nearly 60 percent of our household energy use goes toward heating. Another 15 percent is consumed in heating domestic water. Cooking and refrigeration are lower on the list, followed by lighting, laundry, and entertainment usage.

There are things you can do in the home, most requiring very small investments, that will contribute to a noticeable decrease in your monthly fuel and power bills and will reduce the amount of greenhouse gases introduced.

Here are some tips.

1. **Beware of power vampires.** Power vampires are electrical devices that continuously draw electricity from power outlets, even when not doing any work. Typically, a household might have a dozen or more devices like this (TVs, VCRs, DVD recorders, battery chargers, and game devices). Home office offenders include computers, cell phone chargers, printers, answering machines, monitors, and copiers. They can be identified by the transformers, which are small black or white power brick connected to the cord. Some devices have internal transformers, but you can identify them because the units will be warm to the touch, even when off.

   Individually, power vampires only consume a few watts, but when considered collectively over the course of an 8,760-hour year, the cost adds up, and the aggregate can be 5 percent or more of the total household electricity budget. The simplest solution is to unplug vampire devices when they're not being used. A better one is to invest in a power strip or antisurge device with six or more outlets, then the whole bar can be turned off with a single switch.

2. **Rethink your equipment.** Computers, monitors, fax machines, printers, and copiers have come a long way since they were introduced. Older desktop computers and their monitors, especially, are legendary power hogs. If you can afford to replace outdated equipment, do so with Energy Star labeled devices (though the rating system is not optimized for home computers). Keep in mind that laptops are about two to three times more efficient in terms of power consumption than desktops, even in newer models. Short of replacing your computer, you can reset your computer's "sleep" mode to reduce its power consumption significantly. Savings can total up to $50 per year.

3. **Become a luminary.** Compact fluorescent bulbs use about 70 percent less energy than incandescents while delivering an equivalent number of lumens (a standard unit of brightness). They also last eight to ten times longer than regular bulbs. The additional cost of CF bulbs is more than paid for by the

savings. Furthermore, the recent introduction of light-emitting-diode lamps, or LEDs, open a new world of energy savings as these consume even less energy than CFLs and can last more than 35,000 hours.

Take advantage of daylighting in places where you need light, such as reading chairs, work areas, and sewing rooms. If window space is at a premium in your home office, you could use lighter colors on the walls to reflect the light you receive.

4. **Stop the draft.** Storm windows are a recognized payback investment in cold climates. Better-quality triple-track units with low infiltration rates and substantial felt-pile weatherstripping between panes are the best choice. Aluminum-frame storms should be installed with caulk against the window frame to reduce drafts. If you're economy-minded, prefabbed storm kits or even clear, 6-mil polyethylene film stapled in place beneath wooden lath strips will be effective in keeping cold air out and warm air in.

5. **Save costs on hot water.** If you use a gas or electric water heater, you can save a portion of your water-heating bill by wrapping the tank in an insulating jacket. Newer heaters should have a satisfactory layer of protection, but older units most likely could use the help. Wraps are made of polyvinyl-covered fiberglass or closed-cell foam and can easily be installed by homeowners.

   Also, lowering the heater tank's thermostat to 110°F (43°C) saves energy. People with electric heaters can install timers that limit the frequency and duration of heating cycles. And consider installing a tankless water heater. These on-demand electric or gas units only heat the water that's actually being used at the moment rather than heating and storing a large tank of water.

6. **Set back and save.** By lowering room temperature 5°F (2.8°C) over an 8- to 12-hour period and resetting it in the daytime, you can save more than 5 percent on your heating bill. Setback thermostats are a great investment; they're programmable and can be set to lower house temperatures according to your schedule.

7. **Chill wisely.** Refrigerators—especially older models—are energy hogs. If you have an energy-saving feature on your refrigerator, use it. The cabinet will be less likely to show condensation in the winter months. Check the temperature inside the unit using an outdoor thermometer; readjust the setting if needed to 38°F (3°C). Keep the ventilation grilles at the bottom and back of the cabinet free of dust and pet hair, and don't shove the box so tightly against the wall that it blocks its own airflow.

CHAPTER 5

BASEMENT AND ATTIC

The basement and attic are buffer zones and catch-all storage spots for whatever doesn't fit in the rest of the house. In some homes, these spaces evolve into valuable bonus rooms: family rooms, bedroom, offices, crafting spaces. Weatherproofing programs emphasize the attic and basement, where so much heat and cold can be lost…or gained. In this chapter, we offer ideas for safe pest control, secure storage, and energy efficiency.

Use Safe and Effective Indoor Pest Control

Cost savings

Between $4.79 and $15 per treatment depending on application

Benefits

Inexpensive indoor pest control without the use of poisons or chemicals

You can rid your home of unwanted pests—fleas, roaches, mosquitos—without using harsh chemicals. *BackHome* contributor Catt Foy discusses effective ways to rid your home of pests without harmful chemicals.

PREVENTION

The major indoor pests are roaches, mice, ants, fleas, ticks, flies, mosquitoes, and moths. One of the primary steps in ridding your home of unwanted pests is prevention. Cleanliness is critical, as dirty dishes and overflowing trash containers provide an enticing banquet to rodents and insects. A clean home will be less attractive to such invaders and will reduce those already in residence. Vacuum thoroughly and frequently, keep food put away, avoid collecting dirty dishes, and keep trash containers closed. Empty litter boxes often, and clean up after pets immediately.

The next step to prevention is outside the home. Since most pests enter the home from the garden or lawn, barriers will go a long way toward keeping your home pest-free. Plant insect-repelling plants close to the house and along foundations, and keep collections of branches, firewood, and compost heaps away from the house.

Pesticides used in the home can linger in carpets and fabrics for months or years, because there is no wind and rain to break them down and flush them away. Pesticide residues in the home can be very toxic to small pets, especially birds.

MOSQUITOES

Mosquitos love standing water, so remove containers around the home that might hold rainwater, and fill holes in trees, driveways, and sidewalks. Free gutters of debris so they will not collect water. Also, consider planting tansy or basil close to the house or in window boxes.

ANTS

Ants are also attracted to water sources, so moisture control is important. Dry dishes immediately, and repair leaky pipes and plumbing fixtures. Ants can get in nearly anywhere, so caulk around windows, doors, and other openings. Smearing petroleum jelly around electrical outlets also discourages ants.

Keep all food preparation areas free of crumbs, and keep countertops dry. When ants are in carpet, vacuum and add talcum powder to the bag. If ants are in houseplants, submerse the pot in warm, soapy water for up to 30 minutes. Add crushed eggshell to potting soil to prevent ants from returning.

One of the most effective ant deterrents is to sprinkle cinnamon or cumin along windowsills, the edges of kitchen counters, and in or under cabinets. Plant peppermint, spearmint, pennyroyal, and/or tansy close to the house or near places where ants are likely to enter.

ROACHES AND MICE

Roaches and mice are also discouraged by a clean kitchen. Prevent mice from entering the home by covering openings with fine mesh screen and caulking all holes (or use cement). Remember vents, ducts, and openings around drains and dryer exhausts. Keep the woodpile, compost heap, and other rodent habitats away from the house. And, of course, a house cat or two can help control mice.

To prevent roaches, put away all food products at night, including pet food bowls, and don't leave water standing in the sink or other open containers. Roaches are especially fond of collections of old newspapers, boxes, and other paper. Recycle newspapers as soon as possible, or seal saved newspapers in large plastic bags or boxes. Keep dry pet food in a closed container.

Roaches, mice, and ants can also be eliminated by using boric acid sprinkled in out-of-the-way places, but boric acid is toxic and can be dangerous to pets and children. An alternative that is just as effective is to make your own roach powder—mix equal parts baking soda and confectioners' (powdered) sugar. The fruit of the Osage orange (hedge apples) also kills roaches. Place the halved fruit in places where roaches are likely to be found.

FLIES

Flies, too, are attracted to garbage and other smelly debris, including animal droppings and compost heaps. Clean up after your animals, keep garbage in closed containers, and place the compost heap away from the house. Make sure that all screens are in good repair. Flies and fruit flies that make it into the house can be lured into a trap. Place white wine, apple cider, or soft fruit in the bottom of a wine bottle or mayonnaise jar. Fit the top with plastic wrap held in place with a rubber band, and poke a few holes in the plastic. Flies and fruit flies will get in but cannot get out.

Another fly-trap method involves making your own flypaper. Boil water, corn syrup, and sugar into a thick syrup. Spread on strips of brown paper bag, and hang in places where flies congregate. You can also hang clusters of cloves, clove sachets, or clove-studded oranges to deter flies.

FLEAS

A teaspoon (5 ml) of vinegar per quart (946 ml) of water in your pet's drinking bowl will help prevent fleas and ticks. Pet expert Caroline Swicegood recommends this natural flea-spray: place orange peel, grapefruit peel, 3 garlic cloves, 1 teaspoon (0.7 g) of rosemary, and 1 pint (475 ml) of water in a blender and puree. Heat the mixture for 15 minutes on low heat. Cool and strain into spray bottles and spray pets thoroughly. To further deter fleas, vacuum frequently and place eucalyptus, rosemary, fennel, or red cedar shavings where your pets sleep.

MOTHS

Repel moths by placing sachets or bags of dried herbs such as mint, lavender, or rosemary. Mint tea bags work just as effectively. Place stored clothing in sealed containers, if possible, or place hanging clothes in plastic bags.

Try No-Nonsense Mildew Control

Cost savings

Around $3.25 per application for cleanup and thousands for prevention

Benefits

The prevention of costly mold remediation, healthy low-cost cleaning options

Stop mold and mildew before the powdery bloom (achoo!) takes over damp, dark spaces in your home—bathrooms, laundry rooms, kitchens, or any room with limited circulation. Treating minor mildew once it has appeared needn't be a risk to your health. You can remove stains and bloom without strong chemicals or bleach. Simply mix ⅓ cup (80 ml) of hydrogen peroxide with 1 cup (235 ml) of water in a spray bottle. Or, use a 2:1 mixture of distilled white vinegar and water. Allow the mixture to soak on the moldy surface, then wipe off with a cloth.

Here are some cleaning strategies for problem areas in the home:

Bathrooms: If there's an exhaust fan in the bathroom, be sure to turn it on before you shower, then leave it on for 10 or 15 minutes after you're done. Exhaust fans should not empty into attic or crawl spaces, where the moisture can do real damage. They should be adequately sized to remove moist air, and should not be compromised by other fans in the house (such as a range hood) that will affect their ability to move indoor air outdoors. (Without a source of replacement air, the stronger fan may "steal" air from the weaker one.)

Shake the shower curtain after use and open it partway to allow humid air to exit the bath. If you have shower doors, get a small squeegee and wipe the glass surface clean before opening the doors. Get in the habit of opening the bathroom door while you run the fan, and then leaving it open to allow the room to dry out.

LAUNDRY ROOM: Make sure that the dryer's vent hose is properly sealed at connections and is not torn or split, and that the outlet isn't blocked or its exhaust flap stuck. Dryer exhaust air is laden with moisture and must be directed outside and not into a garage or crawl space.

Kitchen: Make a habit of running the range hood when baking or using the stovetop. Leaking or sweaty pipes, leaky roofs, and overflowing drain pans in HVAC systems can also be a not-so-obvious source of moisture within the structure of a home.

Best Practices for Battling Mildew

Some basic maintenance around the home can prevent mold and mildew.

IMPROVE CIRCULATION.

In general, rooms with limited circulation tend to be a target for mildew and need to be ventilated. A window fan, or in larger rooms a ceiling fan, will go a long way toward keeping air moving and drying out, and at a minimum of energy cost. In worst-case scenarios, it may be necessary to invest in a dehumidifier, which in the long run is less expensive than professional mold remediation.

CONTROL CONDENSATION.

Condensation is another source of moisture in the home and can be the result of poor insulation, worn weather stripping around doors and windows, or lack of window insulation. An energy and weatherization audit can reveal the source of insulation problems and will help you to determine where whether stripping is ineffective and whether storm windows or other window options might be needed.

PART 2
THOSE SPECIAL INTERESTS THAT MAKE LIFE SPECIAL

CHAPTER 6

PETS: PART OF THE FAMILY

A home isn't complete without a furry friend or two. This chapter shares methods for safe and low-cost flea control, healthful pet food recipes, how-tos for making pet futons, and other goodies. A family could easily spend a fortune on veterinarian care, so tune in to some important guidelines for keeping your pet healthy at home.

Say Farewell to Fleas

Cost savings

About $113.83 over a six-month dosage period based on the cost difference between commercial flea treatment and brewers yeast supplement

Benefits

Avoid using toxic chemicals on your pets and in your home

Fleas can be a difficult problem. They breed incessantly and feed off the blood of both humans and pets. Their bites, while not deadly, can be annoying. They can weaken kittens, as well as sick or older animals. Commercial flea treatments and collars aren't always the best solution, due to their cost and their toxicity to some cats. Here, *BackHome* contributors Joyce and Jim Lavene offer some tips on how to get rid of these pesky pests using far less expensive and more natural approaches.

TREATING THE HOUSE

In one month, a female flea can produce more than 20,000 offspring. The adult flea spends most of its time in carpets and sofas. They bite the closest warm-blooded animal, jump off, and reproduce. Then the cycle begins again. The pupa stage of a flea's development is the toughest to knock out. Even when the adults, eggs, and larvae are dead, the pupa remain, ready to hatch a whole new generation in about two weeks.

Contrary to popular belief, the most important battleground is in your house and yard, rather than the actual pet.

One of the major problems in killing fleas with a pesticide is its recommended frequency of use. Commercial products aren't safe to use more than once a month. Fleas hatch every two weeks. Any battle plan has to take this into consideration.

Here's one effective, inexpensive approach. It's messy, but really not any more so than those powdered room deodorizers. And it kills the fleas. Combine about 12 ounces (340 g) of diatomaceous earth, 2 tablespoons (28 ml) of citronella, and 10 drops of eucalyptus oil can be mixed together and shaken into carpets and furniture every two weeks. You can buy diatomaceous earth at garden supply stores and health food stores.

Vacuum the mixture up after an hour. It also helps to get into the habit of vacuuming more frequently, making sure to get into corners and crevices in the rooms and furniture. Empty the vacuum container after each cleaning and rinse it thoroughly.

TREATING THE YARD

You can buy beneficial flea larvae-eating nematodes from mail order or directly from pet or garden stores. The nematodes attack and kill flea larvae as well as 250 other harmful insects, including roaches and termites. Nematodes are said to be harmless to beneficial insects, birds, and animals.

One variety of nematodes are marketed at garden stores under the brand name Guardian/ Lawn Patrol.

Even if your pets don't go out of the house, you'll want to make sure you yourself aren't bringing in fleas. To do your best at preventing that, treat the yard once a month.

TREATING THE CAT

Good health is the most important weapon in treating cats for fleas. A good balanced diet and nutritional supplements will help a cat fight infestation. You can add brewer's yeast supplements directly to the cat's food. They actually make the animal taste and smell bad to fleas. But use it sparingly to avoid intestinal gas: 25 milligrams per 10 pounds (4.5 kg) of body weight.

You can also buy a small flea comb and use it every couple of days. Combing your cat is an easy way to check for adult fleas. Do this outside in case you find fleas on the comb. If you don't see adult fleas in the coat, use a spray of 1 pint (475 ml) of water mixed with ten drops of orange oil to control the problem between baths.

Hopefully, you'll find that within a few weeks, your house will be free of fleas. It takes a little work and more attention, but it's worth the effort to be able to walk through the house with white socks on and not see any black flea dots. And, your cats will be happier and healthier, too.

TREATING THE DOG

Dogs, as we all know, seem to be a magnet for fleas, a situation which can get out of hand if not controlled properly. Brewer's yeast is a common addition to the diet, but it comes with a caveat: Some dogs are sensitive to it, and it can cause uncomfortable gas or bloating, and even a skin rash in some cases. And it may not have as much effect on fleas as was once thought. Some dog food already has nutritional yeast as an ingredient, so use no more than one teaspoon daily in the food. Another natural repellent is plain apple cider vinegar, which can be added to your pet's drinking water at a dosage of 1 teaspoon (5 ml) per daily water ration.

Bake Dog Biscuit Treats for Pennies

Cost savings

About $22.35, based on a 60-count supply

Benefits

A more healthy less expensive dog treat

Your four-legged companions of the barking variety will appreciate these easy-to-make doggy treats. This recipe was concocted for a fundraiser to aid a local animal shelter, but *BackHome*'s food editor Judy Janes discovered that her four dogs liked them so much that it was far less expensive to make them herself than to consistently buy them. (Although she still makes the occasional purchase to support the cause.)

You can purchase the dog bone-shaped cutter at a kitchen supply store, or you can make one from an empty tuna can. Begin this project by removing the bottom from a can with a hacksaw, taking care to keep the sides from crimping and collapsing while you work by shoving a block cut from a stout piece of wood into the container's open end.

Remove the burrs from the edges with a fine file, then gently bend the sides into an oval shape with gloved hands. You can bring each tin to a pretty fair representation of a bone by using needle-nose pliers to make the indentations at the ends of the bone, forming the rounded corners carefully around a short section of an old brush handle, and squeezing the sides to a concave shape. Use the rolled can edge to push the metal form through the flattened dough when you form the biscuits.

Health-Giving and Delicious Doggy Biscuits

Makes 60 biscuits

- 2 cups (160 g) rolled oats
- ⅔ cup (140 ml) canola oil
- 2 bouillon cubes dissolved in 3 cups (705 ml) hot water
- 1 envelope (3 oz) (85 g) powdered milk
- 1½ cups (205 g) cornmeal
- ½ cup (56 g) wheat germ
- 1 egg, beaten
- 6 cups (750 g) whole wheat flour
- 3 tablespoons (12 g) chopped fresh parsley

Preheat the oven to 325°F (170°C or gas mark 3). Grease two cookie sheets. In a bowl, mix all the ingredients with a dough hook or knead them by hand until they form a large, smooth, very stiff dough. On a lightly floured surface, roll out the dough until about ½ inch (1.3 cm) thick. Cut the biscuits out with dog bone–shaped cookie cutter, and place them close together on the cookie sheets. Keep kneading and rolling out scraps until all of the dough is used. Bake for 1 hour. Turn off the oven and allow the biscuits to cool and dry in the oven. When they are completely cool, place four biscuits to a bag in zip-up bags, and store them in cool place.

Visit the Vet Less by Using Natural Healing

Cost savings

These remedies cost from nothing to a few cents per treatment, saving you expensive trips to the vet

Benefits

Natural health treatments with little expense involved

A pet's illness, or imbalance, is often due to improper diet, poor living conditions, or lack of exercise. Here, contributor Shawn M. Schulz shares the experiences she's gained in trusting natural healing for her pets.

As a child, I learned from my grandfather about natural cures—plants gathered from the land, ways passed on by his own father. I remember going into the cool, dark milk house and seeing shelves lined with bottles and tins. There was homemade liniment for sprained muscles, pine tar for thrush, camphor for caked udders, flowers of sulfur for sore hocks, and a jar of my mother's black drawing salve that was used on every animal and about every person, too.

Our farm was sold long ago, but I still use herbal remedies for my family and pets. It's the way we have always done things, and I have seen the results for too many years to break the tradition now.

SKIN PROBLEMS

The toxins we put on our pets, and the chemicals we feed them, turn up as skin problems. My English mastiff had always been bothered with itchy, flaky skin with hair loss on her rump and tail. Years ago, I placed her on an all-natural vegetarian diet to help her arthritis. In a month her arthritis had improved, and her skin condition had, too.

So look to your pet food first for the source of skin problems. Foods without chemicals, preservatives, or food coloring are available in stores. Also look for pet-food cookbooks, including Joan Harper's *The Healthy Cat and Dog Cook Book*, with natural recipes for cats and dogs.

Skin problems can also be caused by allergies to fleas, insecticides, even aluminum feeding pans. Try switching to stainless steel if you're using aluminum, and change to all-natural flea control if you're using products with chemicals.

FLEAS AND TICKS

Fleas spend most of their lives off your pet. They hitch a ride only when they need a drink of blood. This means that most of the time they're in your pet's bedding and your rugs, draperies, and furniture. Frequent vacuuming, about every two days, will cut down on their population. Just be sure to change and dispose of the vacuum bag, or you'll nurture a flea nursery. Launder your pet's bedding frequently, and dry it in the sun or in the dryer to kill the eggs.

Bathe your pet once a month, or every 2 weeks if you have a bad infestation. This cleanses the skin of toxins as well as fleas. I use a natural castile soap and herbal flea rinse. To make the rinse, add 1 cup (225 g) each of dried rosemary, dried rue, dried wormwood, and dried European pennyroyal to 1 quart (946 ml) of vodka or beer. Keep the rinse in a covered jar, and let it steep 4 days, shaking it well each day. At the end of the fourth day, strain and bottle it. You can add a dram of essential oil; eucalyptus, sweet orange, cedarwood, and citronella oils are offensive to fleas. The final mix can be diluted with up to 2 gallons (7.57 liters) of apple cider vinegar. After bathing my pets, I sponge or pour the rinse over them and don't rinse it off.

If you have a kitten, puppy, or old animal, use a milder rinse. Simply add one lemon or orange rind, slightly bruised, to each pint (473 ml) of boiling water. Let this mix stand overnight, strain it, and sponge the animal with the rinse. Fleas dislike the oils in the orange and lemon rind, so the rinse repels them but doesn't kill them.

Ticks are usually repelled by the same methods. However, if you find one clinging to your pet, take care in removing it. The head usually is embedded in the animal's flesh, and failure to remove the whole tick can cause infection. Putting a drop of pure citronella oil on the tick will make it back off fast.

BREWER'S YEAST: A BASIC

Brewer's yeast helps repel fleas and ticks from the inside out. It can also be used as a flea powder; just dust it on your pet's skin.

A large amount of brewer's yeast in your pet's food should have calcium and magnesium added to it. If you can't find this already mixed, add ½ cup (113 g) of calcium lactate and 2 tablespoons (30 g) of Epsom salts to every pound of yeast. Be sure to get brewer's, not baker's, yeast. My 150-pound (68 kg) mastiff, Ginger, gets a tablespoon of the mix every day; my 20-pound (9 kg) cocker gets a teaspoon. Cats are fussier, so you should start with a sprinkle, slowly working up to ¼ to ½ teaspoon (1 to 2 g) daily.

You can add various things to this mixture, depending on the state of your pet's health. If I suspect a sluggish thyroid, I add ¼ cup (57 g) of sea kelp powder, rich in minerals and vitamins, and supplying iodine for the thyroid gland. For Ginger, who is getting on in years, I add ½ cup (113 g) alfalfa powder to the mix, to keep her arthritis under control.

MANGE

Animals with acutely inflamed skin conditions, such as mange, may benefit from fasting, which is a natural response in animals to illness. Weak or malnourished animals, however, should not fast.

Offer plenty of water during the fast, as well as vegetable broth with a little natural low-salt soy sauce added. My vegetable broth consists of carrots and leftover peelings—any vegetable and other except members of the cabbage family. Dogs should fast for 3 to 7 days and cats for 3 to 5 days. Gradually reintroduce foods, soft gruels at first and then firmer foods, until you're back on a new, improved diet. Limit milk and oils until the mange clears.

Externally, fresh lemon juice or strong black tea patted on sore spots may help. Dilute the lemon juice with water if it seems too strong for the animal's skin, Bathe the pet weekly with castile soap until the condition clears up.

RINGWORM

Ringworm, like mange, indicates the animal's general health is not up to par. Clip the hair around the bare spot and about ½ inch (1.3 cm) beyond to help prevent the spread of the disease. Carefully get rid of the hair, since ringworm is contagious. Wash the animal's bedding and dishes often in hot soapy water.

From the common plantain, you can make a tea that will help rid your pet of ringworm. Chop the leaves and root, obtaining ¼ cup (58 g) of plantain for each cup (235 ml) of water. Boil the water, and steep the plantain for 10 minutes. Cool the liquid, and massage it onto your pet's skin twice a day until the condition clears.

I have also had success using oil of lavender to paint the spots. Be sure to get the pure essential oil, not the fragrance. If the oil seems too harsh for your pet's skin, dilute it with a few drops of cooking oil. Raw lemon juice is also effective, as it seals off air from the ringworm.

Stitch Up a Fab Futon for Fido

Cost savings

Between $36.11 and $71.06, depending upon the store-bought source

Benefits

Just as attractive as store-bought and the opportunity to choose your preferred fabric and pattern

Most of us have a soft spot in our hearts for our canine companions, and *BackHome* contributor June Higgins is no exception. In fact, June has a soft spot on her living room floor, and she put it there—for Beedj, her aged collie dog, when she made this truly inexpensive pet bed.

Old Beedj, like most dogs who've seen their share of sunsets, has a certain fondness for—well, rest. Give him a loafing place warm and soft—the good living-room sofa, for instance—and he'll return to it at every opportunity, no matter how often you scold him. And once he's landed, yawned, and settled in with a stretch, sigh and a sleepy smacking of chops, nothing short of a crowbar will rouse him from his doggy doze.

Time to apply a little canine discipline, you say? Well, maybe. But hey—do *you* like sleeping on a floor? We decided a better approach would be to get Old Beedj a bed of his own—a bed so warm and comfortable our furniture would seem rock-filled by comparison.

The idea sounded good, until we looked at the prices of quality dog beds in pet stores and mail-order catalogs. We weren't willing to pay that kind of money—so instead, we made one ourselves. Here's how we did it, and by adjusting the sizes to suit your dog, you can do the same. The grand total for our materials was $8.89, including tax, thanks to using fabric remnants and leftover egg-crate foam from the fabric store.

MATERIALS

- Egg-crate foam, 42 x 32-inch (107 x 81 cm) piece (ask for a leftover; ours cost less than $5)
- Fabric remnants to cover foam—45-inch (114 cm)
- Zipper—30-inch (76 cm)

STEPS

Cut fabric pieces. Cut fabric into two 43 x 33-inch (109 x 84 cm) rectangles. This allows for ½-inch (1.3 cm) seams all around.

Position the zipper. Placing the two pieces together with right sides facing, machine baste a seam along one 33-inch (84 cm) side with the largest stitch available, then press open the seam. With the material open flat, and right side down, position the zipper, also right side down, over the seam, and pin it in place (a). After stitching the zipper by machine, carefully remove the basting stitches and unzip the zipper.

Finish sewing. Place the two pieces of fabric back together wrong side out and stitch ½-inch (1.3 cm) seams the rest of the way around the case (b). Stitch diagonally across the corners as far in from each corner as the thickness of the foam (c). Turn the bed right side out.

Fit egg-crate foam. Fold the foam in half and wedge it inside the bed. Smooth out the foam, and secure it by zipping the bed shut.

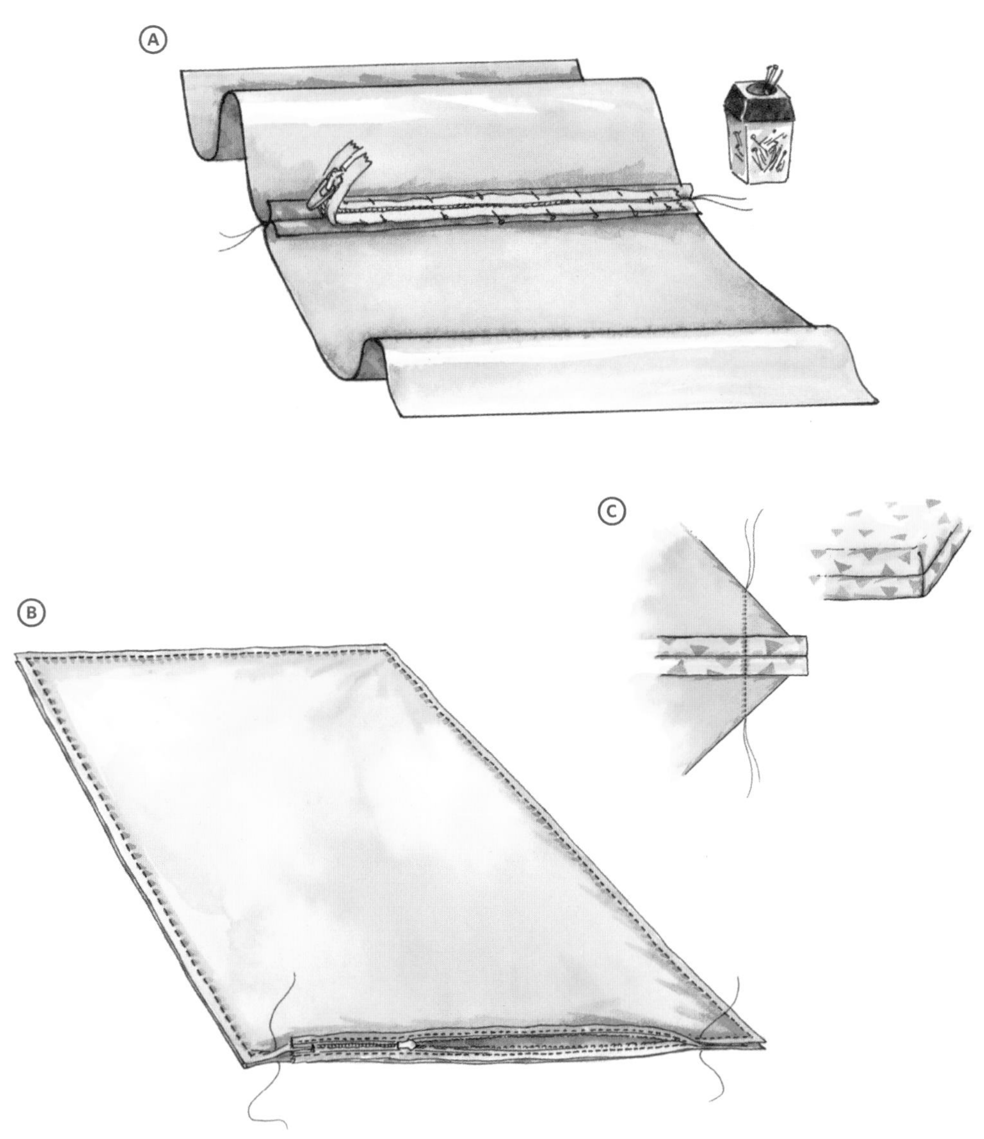

CHAPTER 7

CRAFTS: DO IT YOURSELF

Crafts of all kinds can provide thoughtful gifts for loved ones. This chapter covers a range of useful homemade goodies that are inexpensive and impressive. For the entrepreneurial spirited reader, these ideas could offer another revenue stream through selling handmade items at craft fairs. Consider the possibilities!

Make Recycled Sandals from Rubber Tires

Cost savings

Between $29 and $90

Benefits

A custom-fit sandal that will last better than most store-bought types

Do-it-yourself skills are always a valuable addition to your personal toolbox, and contributor Thomas J. Elpel makes his living teaching skills to students at Hollowtop Outdoor Primitive School in Montana. There, he and his wife, Renee, run the school and a general store. Thomas's brand of self-sufficiency isn't limited to primitive living, but it can apply to the rest of us looking to save money by making our own footwear.

I'm hard on shoes. It's not uncommon for me to go through half a dozen pairs, or more, each year. Some time ago, my friend Jack Fee and I were preparing to go out on a three-week expedition in the mountains. He made a new backpack for the trip, and I made some new moccasins. Jack told me a story about Indians from Mexico coming to the United States and winning footraces in shoes cut from tires.

Jack had never seen the tire shoes that were reportedly used by the Indians but decided to see what he could do. I have to say I was impressed with the final product, a sort of Teva-style sandal that used no glue and had no stitching or strapping on the bottom of the sole where they would be exposed to the ground. Instead he cut the sole and some side tabs in one contiguous piece out of the tire. The first model was a little crude looking but amazingly comfortable. I decided I, too, would make a pair.

The field tests of our shoes were exciting. We combined the tire shoes with ankle-high moccasins in a sort of "modular" shoe. We wore both the moccasins and the soles when hiking, and then just one or the other around camp. We could use only the moccasins for stalking or only the tires for walking in water. We climbed 10,000-foot (3 km) peaks twice and generally put on miles. I did not wear socks and never washed my moccasins, but my feet were in healthy condition for the duration of the trip—a first for me.

Since then I have developed the idea some more, into the tire shoes here. The most significant modification was the addition of the tab at the back. That tab is not normally necessary, except in water, but without it your feet tend to slide forward off the front of the soles when the tires are wet. With the back tab, your foot is held in place. I also added the rubber buckles and did away with the rope and buckskin ties of our early models.

TREADS AND TOOLS

When choosing raw material, I recommend truck tires rather than car tires. The "corner" of any tire, where the sidewalls and tread come together, is much thicker than the rest. You can work with that thickness in the tabs of the shoes but not in the sole itself. Pickup tires are typically wide enough to work with, and you can make about three pairs of shoes from one tire.

Most important, never use tires with steel cables running through them. All tires have some kind of fibrous reinforcement, typically nylon or rayon threads. Most newer tires also have a layer of steel cables, which is not workable. Still, there are a few billion of the older tires around without steel cables, so you shouldn't have to look far to find some. Just look on the sidewalls of the tire for a printed notice of how many plies of nylon, rayon, or steel are imbedded in the rubber.

We used simple utility knives to cut out our first shoes. Doing it this way you can trace the pattern on the outside of the tire and start cutting. However, this is laborious and not much fun. Indeed, it is hard work, and you could easily slip and cut yourself with the utility knife. Along the way I discovered it is easier and more enjoyable to cut tires using sharp wood chisels or a bandsaw. I've done separate tests, cutting out the shoes with chisels and with a bandsaw, and the bandsaw method is only a little faster. A good set of wood chisels works fine if you don't have a bandsaw.

To use a chisel or bandsaw, though, you must first remove a section of tire. This allows you to run the piece through the bandsaw, or to put it on a wooden block where you can chisel from the inside out.

A circular saw works fairly well for cutting tires into usable sections, except that it creates a lot of blue-black smoke and binds frequently. Any piece should be at least a half inch longer than your pattern. Save as much of the sidewalls as you can, as these are useful later for making the buckles. Do not try cutting through the inner edge of the tire, which has an imbedded steel band to fit the tire snug against the rim.

I suggest making only one shoe at a time and completing it. Finish the one and try it on; you might think of some modifications to improve the next. Few of my pairs of shoes are exactly identical, as I usually find some idea to try on that second shoe.

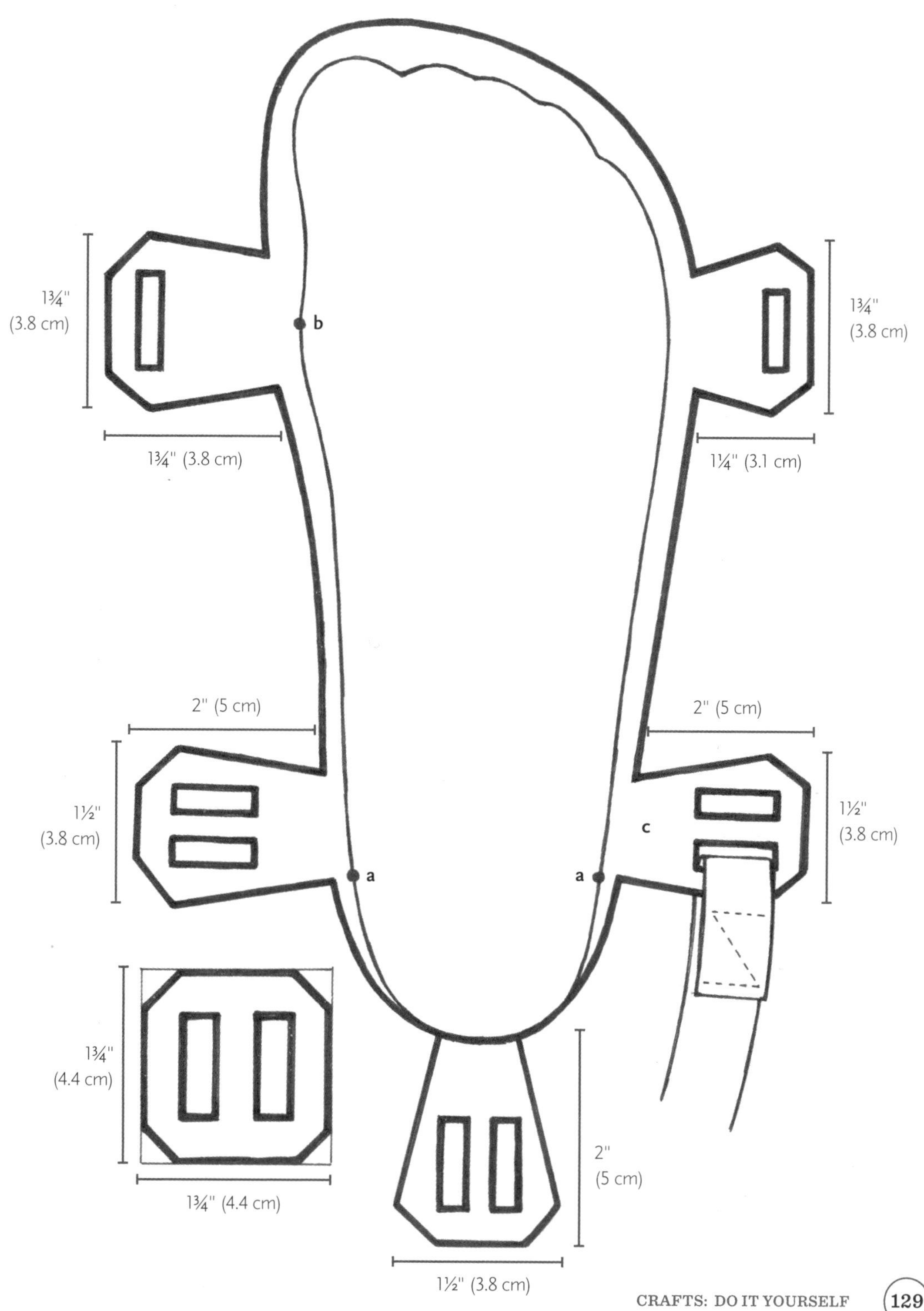
1¾"
(3.8 cm)
b
1¾"
(3.8 cm)
1¾" (3.8 cm)
1¼" (3.1 cm)
2" (5 cm)
2" (5 cm)
1½"
(3.8 cm)
c
1½"
(3.8 cm)
a
a
1¾"
(4.4 cm)
1¾" (4.4 cm)
2"
(5 cm)
1½" (3.8 cm)

MAKING FOOTPRINTS

Plan to spend most of an entire day making your first pair of tire shoes. You'll get faster as you make more.

Start by placing either foot in the center of a large piece of paper, at least 8 1/2 x 14 inches (21.5 x 35.5 cm). Trace around your foot, being careful to keep the pencil straight up and down. Next, make a mark on each side, directly below the point of bone on each side of your ankle (marked a on the pattern). Make a mark at the widest point along the inside of your foot, directly back from your big toe (b).

Remove your foot from the pattern, and sketch a bigger outline around the tracing of your foot. Add about 3/8 inch (9 mm) for the toes and sides but not to the back. Then, using a ruler, bisect the pattern lengthwise, extending the line 3 inches (7.6 cm) past the heel. This serves as a guide to help you sketch the rear tab accurately. Now connect the marks you made by your ankles (a), extending a line 3 inches (7.6 cm) beyond each side of the pattern. The side tabs will be sketched in front of this line. Also draw a line for the front tabs, extending from the single mark (b) across the pattern, perpendicular to the line that bisects the foot lengthwise.

The positioning of all the tabs is quite variable, and you can choose to move them forward or back or at angles to one another; most such adjustments usually work, although the arrangement I have suggested may work more consistently. Problems usually arise with the front set of tabs. When at angles across the pattern they can twist a little and dig into your foot. If the tabs are moved forward or back, the edges can dig into that point b on the inside of your foot. That spot is more pronounced on some people's feet than it is on others'.

Now sketch in the five tabs, as shown on the pattern. These tabs are sized widthwise for 3/4-inch (1.9 cm) wide strapping and should be made according to the approximate dimensions I've shown on the pattern, regardless of how big or small the foot. If anything, you might make some adjustments lengthwise, adjusting for particularly large or small feet. Finally, sketch in the holes that you will cut out to thread the strapping through. This just helps you remember to cut them the right direction when you get to that stage. Cut the pattern out, and (flipping it for the opposite foot) it can be used for both shoes, assuming your feet are fairly similar to one another.

CUT AND THIN

Now, trace the pattern on the inside of the tire, being certain that the pattern is centered and straight on the tire. Even a slight 1/2-inch (1.3 cm) angle along the length of a shoe can cause problems when you wear it.

The next step, after cutting out the shoe, is to thin the four side tabs. The tabs are generally cut from that "corner" on the tire where there is a thick lump of tread. These are easiest to thin on a bandsaw. You can, however, do a crude but adequate job by cutting the lump down with some careful chiseling or with a sharp knife. Thin down as close as you can to the nylon/rayon plies, without actually cutting any of them. This step is not easy by any method I have found, and I typically leave 1/8 to 1/4 inch (3 to 6 mm) of rubber covering the plies, for a total thickness of up to 1/2 inch (1.3 cm). That is still quite thick, but thin enough to work.

Now, to make the tabs flex upward, use a razor blade to slice straight into the tread of the tire at the joint where the tab attaches. Slice in all the way until the plies inside are exposed. Be careful not to cut into those fibers.

Chisel out each of the eyelets, where the strapping will be threaded through. For this I use a 1-inch (2.5 cm) chisel and a 1/4-inch (6 mm) chisel. Be careful not to cut too close to the edge. If you break out the side of a tab, then you generally have to start all over. Also cut a set of buckles (see the photo and illustration) from the sidewalls of the tire. These are easy to do.

FEET TO BE TIED

For strapping, I use a sort of a nylon harness strapping, available at farm and ranch supply stores. Strapping 3/4 inch (1.9 cm) wide works well with the 1-inch (2.5 cm) slots. Cut pieces that are extra long; you can trim them off after you thread them through. Use a match to melt the end of the nylon strap to secure the threads. Starting with the back strap, thread it through the hole marked point c on the pattern, and stitch an inch (2.5 cm) or so of the strap back on itself. Then thread around through the other eyelets, through the buckle, through the other hole on the first tab, and once again through the buckle.

The front strap should be threaded through the buckle, through both eyelets, and back through the buckle again. This system is a little hard to adjust, but once set, I find I can slip my foot in and out without having to tighten or loosen the straps.

The finished shoes should be comfortable to wear, although you may need to do some fine-tuning to get them right. For any serious hiking you should wear a couple of pairs of heavy socks, or moccasins, or bring along some moleskin.

Pound Flowers

Cost savings

Several dollars to $25 or more, in replacing purchased gifts

Benefits

A creative, low-cost artistic pastime and gift source

If you're looking for an out-of-the ordinary artistic pastime that's easy to do and doesn't require a big investment in materials, then consider flower pounding. Other than a hammer, some cloth, and a table to work on, all it takes is a bit of creative eye for color and arrangement and a pleasing selection of plants.

Hammered art can be used to make wall hangings, framed pictures, or collages that reflect the nuances of nature. You can also make note cards on watercolor paper to naturally personalize your correspondence. *BackHome* co-contributors Becky Flatau and Louise Schotz explain the process simply here.

During the winter, you can use specimens such as houseplants, evergreens, plants from the florist, and seasonal plants such as holly, ivy, and poinsettia. During the spring, summer, and fall, the floral possibilities are almost endless. Grasses, flowers, buds, and leaves all lend themselves to a wide variety of combinations.

PREPARE THE FABRIC

Begin with 2 yards (1.8 m) of fabric. Use 200 thread-count cotton material, either white or unbleached. Wash the cloth in hot water with laundry detergent and one teaspoon of Arm and Hammer Super Washing Soda. Wash the fabric again in hot water with no detergent.

Then, put 1½ quarts (1.5 L) of hot water in a 1-gallon (3.8 L) bucket and add ½ cup (113 g) of alum, available from your grocer's spice aisle. Add the fabric and thoroughly saturate it. Pull the fabric to the side and add to the bucket ½ cup (125 ml) of water and 4½ teaspoons (22 g) of washing soda. Leave the fabric in the liquid for 8 hours. When you remove it, do not rinse. Wring it out and line-dry until it's just damp. Then press it with a hot iron while it is still slightly moist.

While you're soaking the fabric, gather your tools and materials.

GATHER YOUR TOOLS

You'll need a small smooth-faced hammer, several different widths of masking tape, a pair of scissors, a set of tweezers to move the smaller stems and leaves, and a piece of hardwood approximately 1 x 8 x 24 inches (2.5 x 20 x 61 cm) for a pounding surface. Don't try to get by with a pine board here, because it'll be too soft.

Of course, you can't forget the main ingredient—a small collection of leaves and flowers, and stems and grasses if you wish.

THE POUNDING PROCESS

First, cut two pieces of the treated and ironed fabric—one for practice and one the desired size for your project. Then prepare to do a practice pound to determine which plants will give the best color. First, lay the chosen plants on the fabric without overlapping them. Then cover the plant(s) completely with tape. Try to avoid overlapping the pieces of tape so that each piece adheres directly to the fabric.

Next, turn the fabric tape side down onto the board so you'll be able to hammer the fabric itself; do not pound on the tape side. Pound each plant until you see the color appear on the fabric. Use a light touch at first, because the plant juices can spatter. As the cloth absorbs the juices, you can hammer with more vigor.

Once you're finished pounding, remove the tape and any plant residue from the fabric. Let it dry. Additional plant material can be applied to the fabric in steps until your work of art is complete, but be sure to dry the fabric between the application of each plant layer.

After you've finished and your fabric has dried completely, you might want to use a black Sakura Pigma Micron archival ink pen (0.25 mm or less) to delineate the shades of color and the shape of the plant. This part is delicate work, accomplished by *very lightly* sketching in short curved lines with the pen.

Hammered art can also be produced on watercolor paper rather than fabric, using similar techniques. To do this, arrange your plant materials on the paper, but do not tape them in place. Cover the plants with a paper towel. Pound the paper while holding the materials in place. Be careful not to hit your fingers!

DISPLAYING YOUR WORK

You can frame your pounded art with wood, sticks, or twigs. Adding beadwork, embroidery, or quilting can also help embellish the picture. Some artists use fabrics to frame the hammered art, or mount it on colored framing matte to complement the colors in the work.

Keep in mind, plant pigments will sometimes fade over time. To diminish that effect, you can use all-purpose inks to shade and add color to your picture before framing it. Fantastix color tool brush points work beautifully when dipped into ink to regenerate color. You must use a very light touch and a very even stroke to do this inking. Use a test piece to practice on and have blotting paper handy to achieve the best results. You must blot every time you re-ink. Don't use a paint brush because it will cause the colors to run together or bleed.

CARE AND MAINTENANCE

Your finished artwork cannot be washed or hung in direct sunlight, because the colors will eventually fade. You might want to treat the finished piece with a UV or fabric protector to preserve the colors as much as possible. When ironing the fabric, take care to always use a pressing cloth.

We've found flower pounding to be a relaxing and resourceful pastime, and one that seems to appeal to adults and children alike. Foraging for flowers and ferns is a nice way to spend part of a day, and working with cloth and creative patterns has its own allure, part of which is the eye-catching outcome of your efforts.

Make Fireplace Bellows

Cost savings

Between $15 and $25 by creating them yourself from scrap materials

Benefits

Saving on kindling and fire igniting fuels, clean and safe

These days, with firestarters and pre-packaged fatback kindling commonplace, the role of the old-time fireplace bellows has slipped a notch or two. But if you get a chance to work the handles and feel the steady draft of air from the nozzle, you quickly realize that hand bellows still have a place by the hearth.

Here is a piece of traditional technology that not only is appropriate, but couldn't be simpler. It consists of two slabs of spade-shaped pine joined to a hollow block with a leather hinge at the narrow end, and sealed everywhere else with a skin of hide loose enough to permit a gentle pumping movement. A flap of material tacked over a couple vent holes in one of the slabs serves as a one-way valve that allows air to be sucked into the bellows chamber from every opening, but permits it to exit from only the nozzle when the bellows is squeezed shut.

MATERIALS

- ¾-inch (1.9 cm) pine board measuring 8 inches (20.3 cm) wide by 30 inches (76.2 cm) in length; a piece of 1 x 10 shelving material is perfect
- 1¾ x 1¾ x 4-inch (4.4 x 4.4 x 10.2 cm) block of pine, birch, or maple (a banister or corner spindle from your builder's supply works well)
- ¾ x 1 inch (1.9 x 2.5 cm) brass hinge
- An old leather shoelace
- A piece of scrap suede, supple leather, or Naugahyde measuring 8 x 32 inches (20.3 x 81.3 cm)
- 5 to 6 dozen ⅝-inch (1.6 cm) decorative antique escutcheon pins
- Contact or liquid cement
- Hammer
- Tape measure
- Utility knife
- Flat file
- Jigsaw (or a coping saw if you'll work by hand)
- Drill; ¼-inch (6 mm) bit and ⅝-inch (1.6 cm) auger-point woodboring bit
- Medium and find sandpaper; wood stain; polyurethane coating for finishing

STEPS

Cut boards. Cut your 3/4-inch (1.9 cm) board into two 15-inch (38.1 cm) lengths. Use the grid pattern provided on page 136; scale the paddle outline onto a large piece of paper so the image measures 8 x 15 inches (20.3 x 38.1 cm). Trace the full-size outline onto one of the boards and cut the shape with a jigsaw. This piece will serve as a pattern for cutting the second board. Round the edges of each handle on one face with a wood file to remove the sharp edge. Finish by smoothing file cuts with coarse-grit sandpaper.

Make vent holes. Measure 6 inches (15.2 cm) from the end of the handle on one of the pieces and bore two 5/8-inch (1.6 cm) vent holes, centered 3/4 inch (1.9 cm) from the middle of the board. The centerpoints of the holes should be 1 1/2 inch (3.8 cm) apart. Next, drill two 1/4 inch (6 mm) holes about 1 1/2 inch (3.8 cm) from the same handle end and 3/4 inch (1.9 cm) apart.

Turn the board over so the inside surface (with the unrounded edges) is facing up. Use fine sandpaper to smooth a 4 x 4 inch (10.2 x 10.2 cm) area around the two 5/8- inch (1.6 cm) holes.

Next, cut a 3 x 4-inch (7.6 x 10.2 cm) scrap from the corner of your fabric material and tack it squarely over the holes so the 4-inch (10.2 cm) edge is parallel with the openings. The fabric flap must be tacked flat along the upper edge only, finished surface down, using three or four brads. This will allow it to open and close as it's supposed to.

Build the nozzle block. Take the banister spindle and cut it down so the block part is 1 1/2 inches (3.8 cm) long and the round part 2 1/2 inches (6.4 cm) long. Using an auger-point boring bit, bore a 5/8-inch (1.6 cm) hole longwise through its center. Sand the inside faces and nozzle block with medium-grit paper, then with fine grit paper. Stain the wood as desired. Once dry, apply two coats of polyurethane. Do not stain or coat the edges of the paddles (except for handles) because the adhesive will stick better to raw wood.

Cut and hem fabric. While the polyurethane finish is drying, cut the fabric to join the wooden parts. Using a utility knife, slice a 2 x 8-inch (5.1 x 20.3 cm) strip from one end of the piece. Then scale up the bellows outline provided (see page 136) in the second grid to create an oblong piece 32 inches (81.3 cm) long and 6 1/2 inches (16.5 cm) wide at the center, gradually tapering to 2 3/4 inches (7 cm) at the ends.

Once the fabric is cut, make a 1/2-inch (1.3 cm) hem by folding the long, unfinished edges and gluing with contact cement.

Attach fabric to wood. Using a hinge and a few No. 4 x 5/8-inch wood screws, join the vented paddle to one side of the nozzle block. Apply adhesive to the final 1/2 inch (1.3 cm) of the block on each side of the hinged face, and to the edges of both paddles. Now, glue on the fabric. Beginning at one side of the block, glue one end of the bellows fabric to the wood, then position the remaining paddle in line with the hinged one, and continue to press the material in place along both glued edges. Work toward the handles, keeping the fabric flat and lining it up with the paddle edges as you go. After one side is finished, secure the hems by driving the escutcheon pins squarely into the edges, beginning at a point 1 3/4 inches (4.5 cm) from the nozzle end and maintaining a space of 1 inch (2.5 cm) between each.

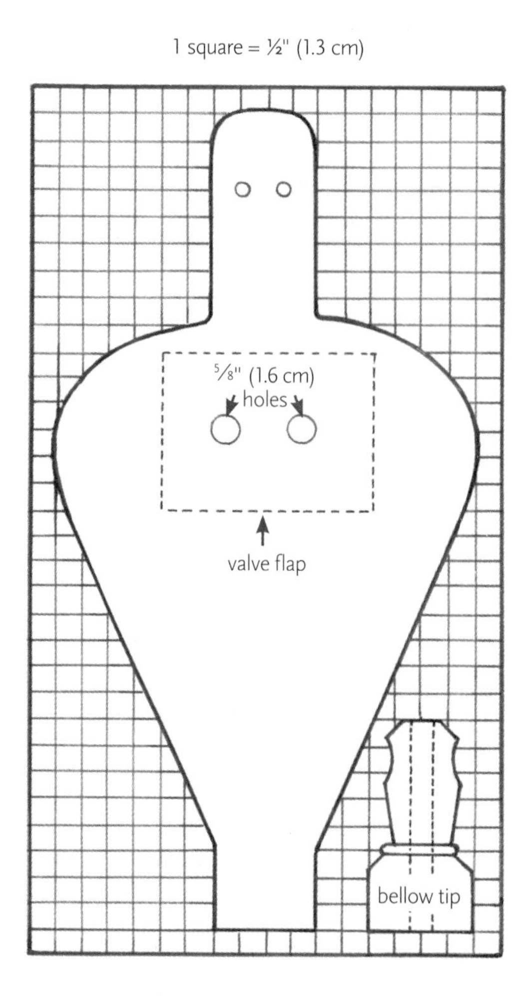

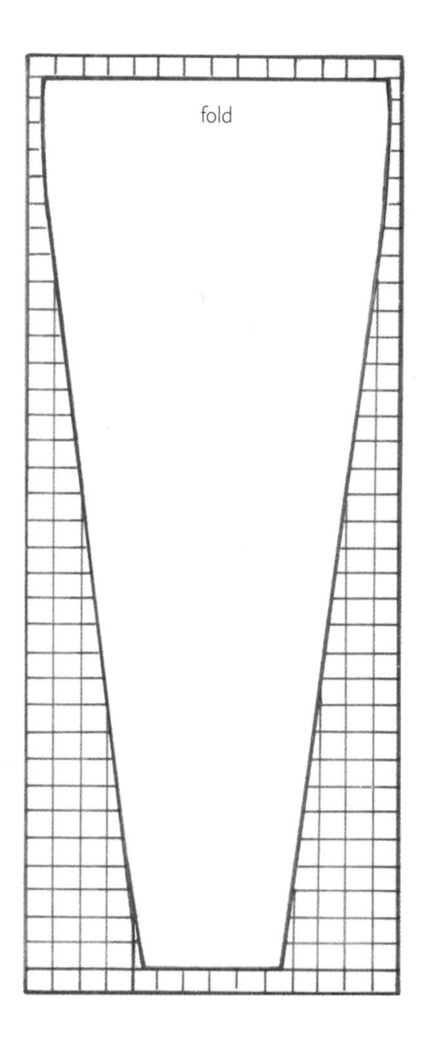

Complete the seal at the inside faces of the handles with more glue, and secure the fabric with a single tack. Then continue around the second half, fastening the fabric with the pins as before. Trim the tail end of the bellows material so it's even with the first.

Beginning at a point along one edge, glue the 2 x 8-inch (5.1 x 20.3 cm) fabric strip evenly between one paddle and the block. Wrap it tightly around the narrow part of the bellows and glue as you go.

Hold the flexible collar in place with two pins along each edge and two centered in each face. String the leather shoelace through the holes in the handle and tie the ends to make a hanging strap.

Make a Simple Solar Cooker for Kids

Cost savings

About 21 cents, by using the sun instead of electricity or gas in the kitchen oven

Benefits

Teaching children the practical value of solar energy and following a plan

Build a solar oven with your children and discover the fun of making simple, delicious recipes. Here, contributor Garth Sundem describes how to make a simple solar oven and offers a delicious recipe too.

There are all sorts of plans out there for making solar ovens—some use pizza boxes, others are more complex. These plans will satisfy the DIY family without overwhelming the kids—the project takes about 2 hours and is appropriate for ages 7 and older (with parental guidance). The oven will heat only to 140°F (60°C) on a sunny day. But you can add features to get more sunlight into the box and stop heat from leaking out. Use the plans as a guide, but let your kids offer suggestions. With adjustments, your solar oven should easily reach 300°F (150°C).

STEP 1: Build the basic reflective box. First, decide how big a cardboard box you will use. A shallower box will make a hotter oven, as the collected sun will have less wall area over which to lose heat. However, a tray with a large surface-area top for collecting sun will gain more solar energy. Cut down the sides of your cardboard box until it is only slightly deeper than the biggest pot or pan you will put in it. If you plan to cook only the muffins described later, your box can be as shallow as 4 inches (10 cm). Remove excess cardboard so you are left with a clean box, open on the top.

The corners and base of this box must be airtight. Seal the outsides with glue and tape if necessary.

STEP 2: Glue foil to the inside of the box. Foil should cover all interior cardboard surfaces—crinkle the foil as little as possible. Paint the foil on the bottom of the box black to retain heat.

STEP 3: Cover the box top. Stretch plastic wrap over the top of the box. Don't worry about neatness because you will be replacing the plastic wrap with a more permanent top when you experiment with this initial design. Congratulations, you're finished (at least that's what you'll tell your kids at this point). You have a reflective box to concentrate the sun, and an enclosed space to trap the heat. Test your solar creation by "cooking" an oven thermometer. Not great, huh? Maybe you reached 140°F (60°C). Now ask your kids what you could do to focus more heat on your box and keep it from escaping.

STEP 4: Explore the options. Encourage your kids to come up with ideas. Address how to insulate the tray. One solution is to "nest" it inside a larger cardboard box, using the dead space between the two boxes as insulation. The exact dimensions of the larger box won't matter as there's at least 1 inch (2.5 cm) of space between its walls and your tray, including height. Fold the top flaps of your larger box closed. Set your tray on the box and trace its outline on the flaps. Cut out the traced rectangle, forming an opening in the flaps the exact size of your tray.

STEP 5: Apply more foil liner. Line the inside of the larger box with foil. Glue the remnants of the flaps closed.

STEP 6: Insulate between boxes. You can use newspaper or old cotton shirts for insulation. Put just enough insulation inside the outer box so that when you set your tray in the cutout opening, the top of your tray is flush with the top of the larger box. Use liberal amounts of glue and/or tape to create a tight seal between your tray and the outer box. Now test the oven outdoors. How much difference did the insulation make?

STEP 7: Prevent heat escape. You stopped heat from leaking out the walls of your oven. Now, how can you stop heat from escaping out the top? How can you focus more sunlight into your oven? For this step, you will need a flat sheet of cardboard, bigger than the top of your larger box. Set your insulating box/tray combo on this larger sheet and trace its outline. Don't cut off the excess! You will fold this excess down on all four sides to create a "lip" that will help this sheet sit snugly on top of your oven. Mark, cut, fold, glue, and tape the top.

STEP 8: Cut an oven opening. The new top blocks the tray opening, which you need so sunlight can flow into the oven. So simply cut out a rectangle the shape of your tray opening. Carefully draw a rectangle that matches the opening to the tray, and cut out only three sides of the rectangle. Fold back the flap. Line the down-pointing side of the flap with foil. This flap will now reflect sunlight into the tray.

STEP 9: Close the gap. Put plastic across the opening of your tight-fitting top, covering the space vacated by the cutout reflective flap. Rather than plastic wrap, use an oven bag or other thick plastic to cover the gap. Cut sections of your plastic bag so that you can stretch a tight cover over the opening on both the top and bottom sides of the solar oven's lid. There is an insulation advantage of having air trapped between the two plastic layers. Be sure to glue and tape the plastic so that it forms an airtight seal.

STEP 10: Test the oven. Prop open the reflector with a metal clothes hanger and point it toward the sun. On a 75°F (24°C) day, your oven should reach near 350°F (180°C).

Grizzly's Huckleberry-Orange Muffins

Grizzly Muffins will cook in solar ovens that reach 275°F (140°) or more. If your oven is a bit cooler, try thinner baked goods, such as walnut/chocolate-chip cookies. Once you gather the ingredients, your seven-year-old will be able to mix the batter with supervision.

- 1 cup (125 g) white flour
- 1 cup (125 g) whole wheat flour
- 1 teaspoon baking powder
- 1 teaspoon baking soda
- 1 teaspoon salt
- 1 teaspoon ground cinnamon
- 2 eggs
- ⅔ cup (100 g) brown sugar

- 1 cup (230 g) yogurt
- 2 teaspoons grated orange peel
- ⅓ cup (75 ml) orange juice
- 2 tablespoons (28 ml) canola oil
- 1 teaspoon vanilla extract
- 1 cup (145 g) huckleberries (or other small berries)
- 1 tablespoon (13 g) granulated sugar
- Butter or paper muffin cups

Preheat the solar oven. In a large bowl, mix the white flour, whole wheat flour, baking powder, baking soda, salt, and cinnamon.

In another large bowl, mix the eggs, brown sugar, yogurt, orange peel, orange juice, oil, and vanilla extract. Add the wet ingredients to the dry ingredients. Mix only until blended.

Sprinkle in the berries. Mix only a little bit, so you don't squish them. Either coat the cups of your muffin pan with butter, or set a muffin cup in each spot. Spoon batter into each cup, filling a little more than half-full. Sprinkle the granulated sugar on top.

If your oven reaches 350°F (180°C), cook for 30 minutes. If your oven reaches only 275° (140°C), cook for 1 hour. To check, stick a wooden pick into the center of a muffin. If any batter sticks to the toothpick, bake for another 10 minutes. Test again. Try to open your solar oven as little as possible. Cool for 5 minutes on a wire rack.

Make a Chair from Bittersweet for Next to Nothing

Cost savings

Between $45 and $75, or the cost of a store-bought chair

Benefits

The satisfaction of designing and creating something unique for your home

Craft projects that use native materials and require minimal use of tools make quick and easy gifts. *BackHome* contributor David McCormick has been making rustic chairs for years. Here, he shares how.

Chairs from bittersweet vines are easy to construct, and you'll find a bounty of material that grows along roadsides and hedgerows. The vines differ in size, shape, and form, so the possibilities are endless. I harvest the vines in late fall, after the leaves have fallen, to better see my prospective choices.

Asiatic bittersweet, not to be mistaken for the American variety, is an invasive import from China found throughout the central, southern, and New England states. The vines envelop shrubs and trees, eventually choking the life out of them. The Oriental species can have blunt thorns, and its leaves are broader than those of the native strain. They are in great supply, and they're easily identified in the fall by their bright red-orange berries. You can harvest these vines with a small pruning saw because green stock up to 2½ inches (6.4 cm) in diameter is all you'll need. Be sure to use living vines, because they are pliable and can easily be formed into various shapes. Cut lengths as long as you can handle in a variety of diameters. You'll trim them to size later.

MATERIALS

- Loppers or pruning saw
- Small cordless or electric drill; ⅛-inch (3 mm) bit
- Phillips head driver
- Steel tape measure
- No. 8 gray or tan decking screws (a selection ranging from 1 to 3½ inches [2.5 to 8.9 cm] in length)

STEPS

STEP 1: Cut wood pieces. Cut two ½-inch (1.3 cm)-diameter pieces to 48-inch (122 cm) lengths, four 2-inch (5 cm)-diameter pieces in graduating lengths from 15 to 18 inches (38 to 46 cm) long, and four 2½-inch (6.4 cm)-diameter pieces to 15 inches (38 cm) long.

STEP 2: Bend and precut. Bend one of the 48-inch (122 cm) pieces into a bow and lay it down flat on your work surface. Holding the bow in place, fasten the four 2-inch (5 cm) diameter sections to the surface of the bow with a 3-inch (7.6 cm) screw at each joint. These become your seat slats. Place the shortest piece near the curved end of the bow, and space each subsequent piece so there's a 2 ½-inch (6.4 cm) gap between all the pieces. Predrill the holes to prevent the wood from splitting, and when driving the screws, don't sink them so forcefully that the heads cause the wood to split.

STEP 3: Position chair legs. Flip the seat over so its top surface faces down. Collect the four $2\frac{1}{2}$ x 15-inch (38 cm) pieces and place them upright against the outside of the bow for legs. Set two front legs between the first two seat slats, and set the two rear legs against the rearmost slat at the open end of the bow. Push leg pieces flat against the work surface, realign them if needed, then drill through each one and into the bow behind it.

Attach each leg with two $3\frac{1}{2}$-inch (8.9 cm) screws. Measure midway between the inside of the rear legs, and cut a section of 2-inch (5 cm) vine stock to match that dimension. Fasten it between the two rear legs using $3\frac{1}{2}$-inch (8.9 cm) screws to make the bottom rung.

STEP 4: Complete the seat. While seat is still flipped, take the remaining 48-inch (122 cm) piece and bend it into a bow shape. Place the curved end of the bow against the underside of the seat bow and the free ends to the inside of the rear legs (against the bottom of the rung). Fasten the curved part to the seat bow using a couple of 2-inch (5 cm) screws. Attach the ends to the legs with $2\frac{1}{2}$-inch (6.4 cm) screws. Flip the assembly over onto its legs.

STEP 5: Make the seat back. Cut one $1\frac{1}{2}$ x 60-inch (4 x 152 cm) length of vine and two $\frac{3}{8}$ x 60-inch (1 x 152 cm) lengths, braiding them together while bending the assembly into a bow shape. Fasten the free ends to the rear slat and rear legs of the chair with $1\frac{1}{2}$-inch (3.8 cm) and $2\frac{1}{2}$-inch (6.4 cm) screws.

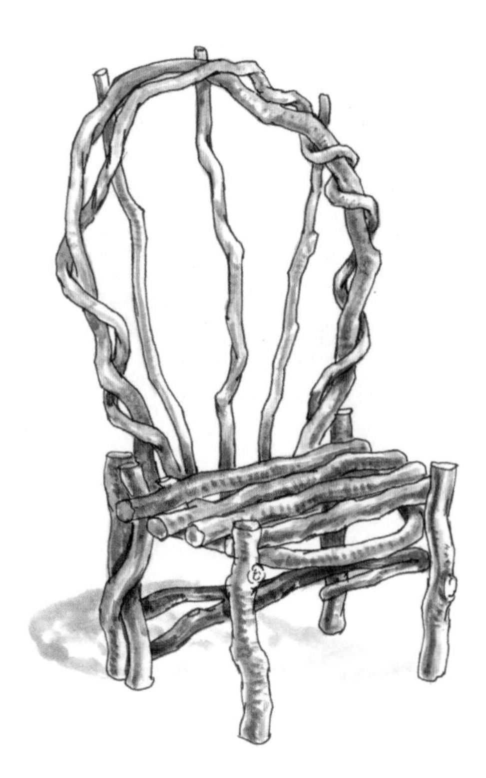

STEP 6: Construct the spine. Weave together two $3/4$-inch (1.9 cm)-diameter lengths of vine (between 32 and 44 inches [81.5 and 112 cm] long) and fasten this to the curve at the back of the chair and the rear seat slat with $1\frac{1}{2}$-inch (3.8 cm) screws.

Make supporting bows by cutting two $1/2$-inch (1.3 cm)-diameter pieces (30 to 48 inches long) and bend them into two small bows. Attach them to the back of the spine and the rear seat slat with 1-inch (2.5 cm) screws. If you prefer spindles, cut three $3/4$-inch (1.3 cm) pieces (between 24 and 38 inches [60 and 95 cm] long) and fasten them in the same way you did the spine.

STEP 7: Add arms (optional). Cut two naturally curved sections of vine long enough to reach from a comfortable height at the center of the seat back to the front. Secure them to the frames with $1\frac{1}{2}$-inch (3.8 cm) screws.

STEP 8: Trim excess vines (optional). Clip excess vines using loppers, or leave the excess for character. Apply clear lacquer for waterproofing if you'll use the chair outdoors.

Make Dyes from Nature

Cost savings

All materials except for soap and vinegar are naturally "found" items, saving several dollars per treatment

Benefits

Traditional naturally dyed wool and woolen goods

The first synthetic dye, a purple color, was developed in the 16th century and was used to make a gown for Queen Elizabeth I. Until then, our ancestors brought color into their lives using dyes derived from plants, bark, nuts, and even insects. Here, *BackHome* contributor Nancy Hamilton offers an introduction to a centuries-old tradition.

Over the centuries, professional dyers using natural substances developed a tremendous amount of knowledge and expertise and could create almost any color imaginable. The dyers belonged to guilds in which only the members shared knowledge, and they guarded their "recipes" closely. All that started changing about 150 years ago, with the manufacturing of synthetic dyes. Today, myriad colors are available from these manufactured dyes, but the trend, particularly among those who use or make their own yarns, is to honor the integrity of the materials by coloring them with natural dyes.

GIVE NATURE A TRY

Here are three easy natural-dye recipes that utilize black walnuts, black walnut leaves, onion skins, and pokeberries. You'll need a 5-gallon (19 L) pot, a ½-inch (1.3 cm) dowel rod for stirring and lifting the wool from the pot, and an old pair of pantyhose (a nylon stocking or knee-highs also work) for holding the dye source.

A few notes of caution:

1. Never use a container or equipment used for dyeing for cooking purposes again.
2. If possible, do your dyeing outdoors on a hot plate or other heat source where the open air can carry away harmful fumes. If dyeing indoors, use a lid to keep the fumes inside the pot.
3. Use commonsense precautions to prevent burns, especially if you're sharing the process with children.
4. Label and store out of the reach of children all dyes and mordants (used to bind colors to fibers) as some are poisonous.

WHAT TO DYE

You can use natural dyes on fabric, but it's difficult to get the color even. It's better to start with a 100 percent wool yarn that has not been treated for machine washing. You can also dye cotton and more expensive fibers, such as silk, mohair, camel, llama, and many blends of natural fibers. (Since blends take dye differently—some lighter, some darker—you can arrive at some interesting variegations.) A first-time dyer should probably start out with an inexpensive wool that can be used for a small knitting or crochet project, such as a pair of mittens. When dyeing yarn, use the back of a chair to wind the yarn into a skein, then tie with loose figure-eight knots so the color can be absorbed evenly.

When working with wool and water, the most important thing to remember is to *never* change the temperature of the wool quickly. Otherwise, you may end up with something resembling felt.

DYEING TECHNIQUES

The following tips will help you become a successful dyer no matter what dye or fiber you're using:

1. Soft water is best for dyeing purposes. Hard water can be softened by adding a little acetic acid or soap. Rainwater is usually soft unless you live in an area of acid rain.
2. Give fiber space in the dye pot, leaving room for freedom of movement and even coverage.
3. Wet wool or other fiber before putting it into the dye bath. If the fibers resist soaking up the water, add a drop or two of dish detergent.
4. Squeeze liquid out of the fiber instead of wringing it. Wringing stretches and weakens the fibers.
5. There are many factors that can affect the final result, so the color achieved may not be what's expected. Keep good records to help you repeat colors you like or follow a different route next time.

BLACK WALNUT DYE

Black walnuts have been used as a natural dye and as an ink ingredient since Colonial times. The nuts produce excellent browns and blacks; they also contain tannin, which acts as a mordant to bind the color with the fiber.

Collect the nuts in late summer and early fall. By this time the hulls will have brown spots on the surface and some decay will have begun. Wear old gloves when handling the nuts, and rubber gloves when working with the dye. Black walnuts stain easily, and the dye is colorfast. Store the nuts in a dry place until you use them.

When you're ready to dye, put the nuts into an old pair of pantyhose, tying its top and bottom. Place in a 5-gallon (19 L) enamel pot and fill with water. Let the nuts soak for several hours or up to a day.

After soaking the nuts, heat them for 2 hours or more, cooking until the liquid is black. Remove the bag and set aside the nuts for the squirrels. Put in the wet wool and simmer for 1 hour. Let the fiber cool in the dye pot overnight. Remove the wool, wash in mild soap, rinse, and hang to dry. The result should be a rich brown color, though cotton yarn usually comes out lighter. If you would like a lighter brown or a tan, repeat the process with additional fiber in the same dye bath.

POKEBERRY DYE

Pokeberries are gathered from pokeweed (*Phytolacca americana*). They were used extensively by Native Americans to make a berry red dye, a practice that was adopted by European settlers. The result is not, however, very colorfast, though it lasts longer when vinegar is used as a mordant.

Pokeweed can be found at the edge of woods or in open fields. Collect a bucketful of the ripe berries in late August or early September. Mash them with a potato masher or a brick. Put the juice, berries, one pint (475 ml) of vinegar, and enough water to fill a 5-gallon (19 L) pot on a heat source. (For a darker, longer-lasting red, use all vinegar instead of water.)

Heat the liquid to a simmer, and maintain for 1 hour. Let the dye cool, then strain it. Dispose of the solids in a plastic bag. Put in the wet fiber and simmer for 1 hour; boiling will produce a brownish red. Remove the wool, wash it with mild soap, and dry it out of the sun.

ONION SKIN DYE

Onions sold loose (not bagged) are usually denuded of a few dry layers before being put on display, and your grocer may not mind saving you some of the leavings once in a while. Expect a yellow, gold, or orange color, according to the type of onion skins used. Cook them in an iron pot for a moss green dye.

Put the skins in pantyhose, tie off the ends, and boil for 30 minutes. Remove them, put in the wet fiber, and simmer for 1 hour. Remove and wash with mild soap. Rinse well and hang out of the sun to dry. This dye will be fairly colorfast if the fabric is not exposed to sunlight for long periods of time.

Stitch Up a Reversible Apron

Cost savings
Between $16 and $25, depending upon how much scrap material is used

Benefits
A two-sided apron with large pockets and fabric patterns you choose yourself

Here's a handy gift for cooks. This reversible butcher-style apron is fast to whip up on a sewing machine. Choose complementary fabrics or patch together some creative combinations.

MATERIALS

- 1 square yard (80 square cm) each of two contrasting fabrics
- Pattern pieces (see diagram)
- Kraft paper
- Pencil
- Iron
- Fabric scissors
- Sewing pins

STEPS

STEP 1: Draw the pattern. On a flat surface spread out a length of 36 x 18-inch (91 x 46 cm) kraft paper. Draw the pattern pieces on the paper (see the diagram), label them, and cut them out. Lay out the fabric pieces, fold them lengthwise wrong sides together, and press them. On the first piece, place all four pattern pieces as shown in the diagram, and pin and cut. Then repeat on the second piece of fabric, but cut only pattern pieces #1 (body of apron) and #2 (pockets).

STEP 2: Prepare apron pieces. On each pocket, turn in the edges 1/4-inch (6 mm) on all four sides, and press. On the fourth (top) edge, turn over an additional 3/4 inch (1.9 cm), and press. Stitch along this top on each pocket.

On each apron piece, measure 20 inches (51 cm) up from the bottom and 5 inches (12.7 cm) in from the side and make a mark. Pin the two pockets in place so that the right-hand pocket has its upper right corner on the mark and the left-hand pocket has its upper left corner on the mark. When pockets are pinned so they are even, stitch them onto the body of the apron.

STEP 3: Assemble the apron. Place the two contrasting apron pieces right sides together. Stitch all around, starting at one side of the bib. At the top of the bib, leave the two slots (see diagram) for the belt and ties. Trim the corners at the bottom, and cut out small half-moons close to the stitching along the curves (this will make the seams lie flatter). Turn the apron right side out and press. Fold under the two raw edges of the top of the bib, and press.

STEP 4: Make belt ties and strap. Turn the belt ties under 1/4 inch (6 mm), and press all around three sides. Fold in half lengthwise, right sides out, and press. Turn the neck strap under 1/4 inch (6 mm) on both long sides, and press. Fold in half lengthwise, right sides out, and press. Stitch all three of these pieces down their lengths and across the ends.

Insert the belt ties into the allotted slots and stitch the slots closed, reinforcing with a reverse row of stitching. Insert the ends of the neck strap into the far right and left sections of the tucked-under bib edge, and secure with two rows of stitching 1/2 inch (1.3 cm) apart.

CUTTING DIAGRAM

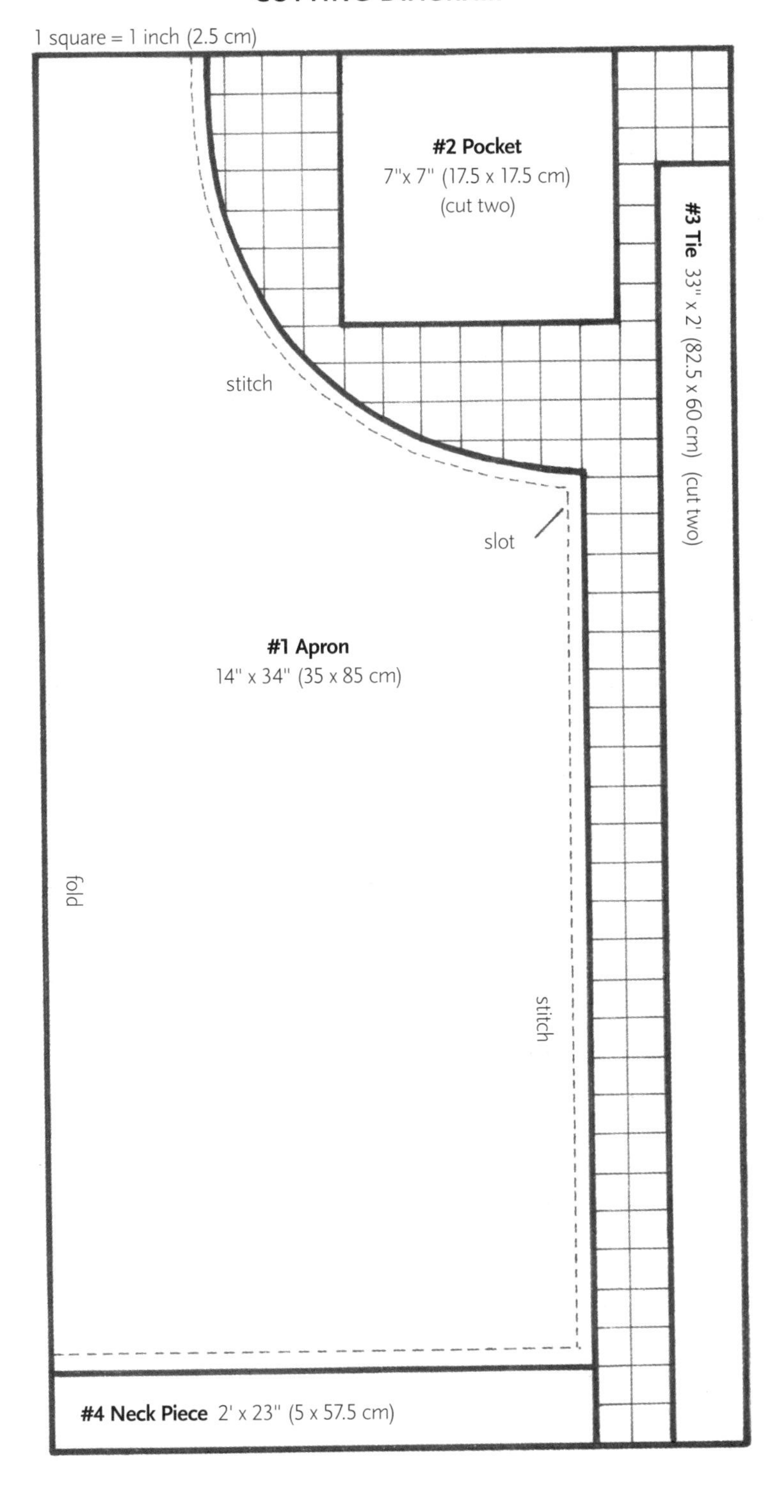

CHAPTER 8

HOLIDAYS: OLD-FASHIONED FUN

Holidays are a great time to bake, decorate, create homemade gifts, and enjoy the spirit of friendship and family. Included in this chapter are more than a dozen ideas for gifts, recipes, snacks, holiday treats, and homemade craft items—all low-cost and fun to make. Holidays are meant to be cherished, and there's no better way to start than saving money!

Make Your Own Holiday Spices

Cost savings

Can be up to half as expensive as store-bought, especially if you buy ingredients in bulk

Benefits

Healthful, fresh spices that you custom-blend yourself

There is nothing like the aroma of a freshly baked pumpkin pie or roasted turkey to put you in a holiday mood. Here, dedicated herb enthusiast and contributor Lynn Smythe offers some traditional blends for your table.

I make my own classic holiday herb and spice blends to use throughout the holidays. Because commercial herb and spice blends often contain salt, fillers, or anticaking agents, I prefer to make my own blends whenever possible.

SPICES

Allspice: The dried, ripe fruits of this plant are used either whole or ground into a powder. Allspice tastes like a blend of cloves, cinnamon, and nutmeg with a slight peppery essence mixed along with the other flavors.

Cinnamon: The dried inner hark of this plant is either used in large pieces, called quills, or ground into a powder.

Clove: Cloves are the dried, unopened flower buds from a tree native to Indonesia. The trees can grow to be 40 feet (12.2 m) tall.

Nutmeg: The dried seed, or nut, of this plant is used in powder form. You can sometimes find whole nutmegs and nutmeg graters available for sale in gourmet stores. Nutmeg used in moderation makes a delightful sweet and spicy addition to your meals.

Sweet Spice Blend

This recipe makes approximately 3 tablespoons (20 g) of sweet spice blend, perfect to use throughout the holidays in apple and pumpkin pie recipes. It also tastes terrific sprinkled on eggnog.

- 2 teaspoons ground cinnamon
- 1 teaspoon ground cloves
- 1 teaspoon ground ginger
- 1 teaspoon ground nutmeg
- 1 teaspoon ground allspice

Mix all spices together in a small bowl. Store in a container with a tightly fitting lid.

Apple Pie with Cheddar Cheese Crust

Crust

- 2 cups (250 g) all-purpose flour
- 1 teaspoon salt
- ¼ teaspoon ground nutmeg
- ⅔ cup (150 g) butter, chilled
- ⅓ cup (75 ml) water
- 1 teaspoon cider vinegar
- 1 egg, separated
- ⅓ cup (40 g) shredded Cheddar cheese

In a mixing bowl, sift together the flour, salt, and nutmeg. Mix in the butter with pastry knife or two knives until the mixture resembles coarse cornmeal. In a small bowl, mix together the water, vinegar, and egg yolk (reserve the white), and add them to the flour mixture. Stir in the shredded cheese.

Use your hands to gently form the dough into a ball. Divide the dough in half. Lightly flour a large wooden board and rolling pin. Roll out a circle of dough approximately ⅛ inch (3 mm) thick. Use a 9-inch (23 cm) pie pan to measure a circle of dough 2 inches (5 cm) larger than pie pan. Place the dough in the bottom of the pan and press well against the sides. Roll out the remaining dough half into a 9-inch (23 cm)-diameter circle and set aside.

Filling

- 6 cups (660 g) peeled, cored, and sliced apples
- 2 tablespoons (28 ml) lemon juice
- 2 teaspoons Sweet Spice Blend (see left)
- 3 tablespoons (23 g) all-purpose flour
- 1 cup (225 g) packed brown sugar

Preheat the oven to 375°F (190°C). Place the apple slices in a large mixing bowl and stir in the lemon juice to coat all the slices. Mix in the spice blend, flour, and brown sugar. Place the filling in the prepared pie shell, and place the reserved pastry on top. By crimping with your fingers, seal together the bottom and top edges of the piecrust and use a knife to trim off any extra. Brush the top of the crust with the reserved egg white. Bake for 55 to 60 minutes, until golden brown. If the edges of the crust begin to brown too much, cover with a ring of aluminum foil.

HERBS

Marjoram: The Greek word *eros* means "mountain," and *ganos* means "joy." In other words, the scientific name of this plant, *Origanum majorana,* means "joy of the mountain."

Rosemary: Rosemary is a symbol of love, friendship, and remembrance. Scholars once utilized its memory-enhancing properties by wearing wreaths of rosemary or carrying sprigs of rosemary to help them when taking exams.

Sage: The scientific name for sage, *Salvia officinalis,* comes from the Latin word *salrere,* which means to cure or be saved. Sage, also know as the herb of immortality, was once thought to be capable of promoting a long and healthy life.

Thyme: Thyme has a variety of attributes associated with it, including increasing one's courage and preventing nightmares.

Poultry Seasoning Blend

Use this to flavor chicken, turkey, duck, or Cornish game hens. It also makes a great addition to stuffing. If you don't have access to fresh herbs, substitute 1 teaspoon of the dried herb for 1 tablespoon (14 g) of fresh.

- 1 teaspoon celery seeds
- 1 tablespoon (4 g) minced fresh marjoram
- 1 tablespoon (4 g) minced fresh sage
- 1 tablespoon (4 g) minced fresh thyme leaves

Mix the spices together in a bowl. Store in a container with a tight-fitting lid.

Make Craft Paper Boxes for Gifts

Cost savings

Between $0.50 and $1.10 per box, depending on its size

Benefits

Really attractive boxes in your choice of colors

The *BackHome* staff tries to come up with a few simple crafts for the holiday season, suitable for adults and children alike. These little gift boxes are so easy to make that once you get the process down pat, you'll find yourself making an abundance of boxes. They're perfect for holding chocolates, little cookies, or jewelry, or hanging as ornaments from the tree.

BOX PREP

First, gather a collection of papers to use. The paper must be sturdier than magazine pages—but magazine covers work well. Collect catalogs, old calendars, greeting cards, or use construction paper.

Before you use your "good" paper, practice with scrap. Make a perfect square by folding a rectangle on the bias to make an isosceles triangle, then trim off the extra. Follow the cutting and folding diagram provided to create the box bottom. Duplicate this pattern to make a box top, increasing the size by 1/4 inch (6 mm) on all sides for a nice fit. (Save the prettiest papers for the tops.)

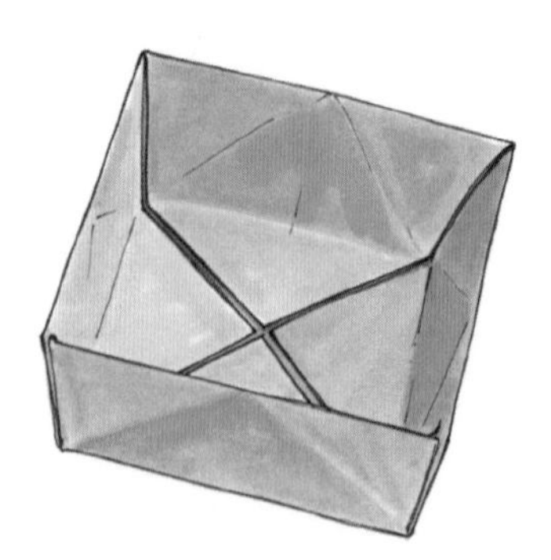

CONSTRUCTING THE BOX

Start with a square piece of paper, decorative side down (a).

STEP 1: Fold all four corners to the center (b).

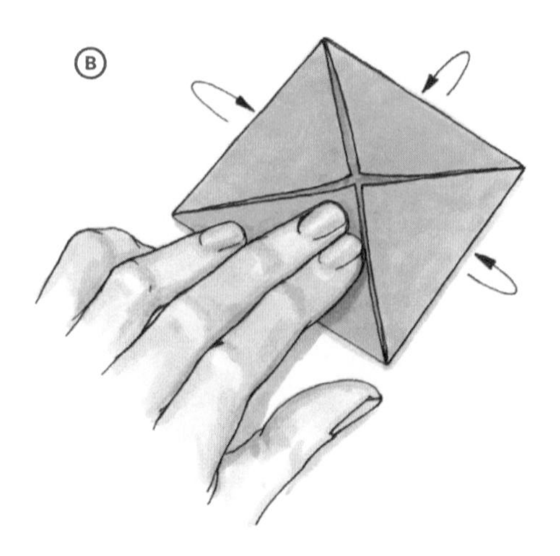

STEP 2: Fold the left and right four sides to the center (c), then unfold. Fold the top and bottom sides to the center, then unfold.

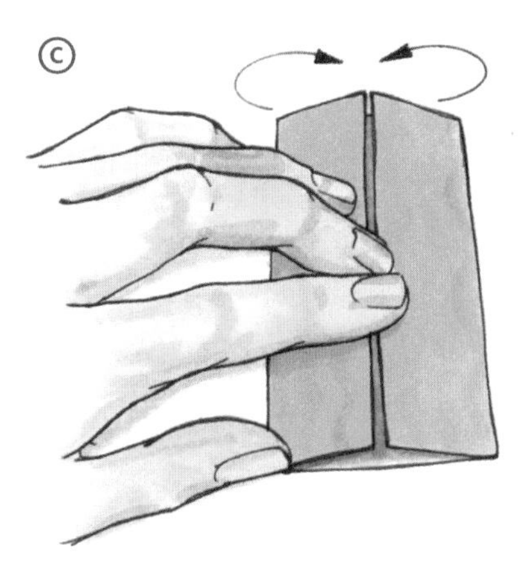

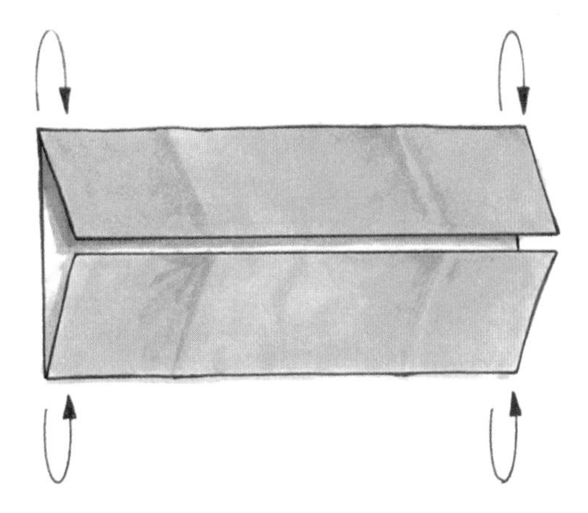

STEP 3: Fold in the squares you made on each corner, making the sides of the box stand up, as shown (d).

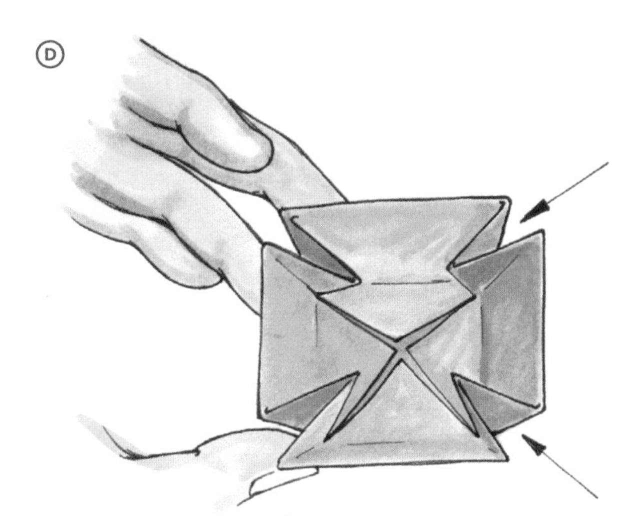

STEP 4: Unfold one side of the box. Press the corner triangles to the unfolded side (e) and refold it down (f). It should hold the shape on its own.

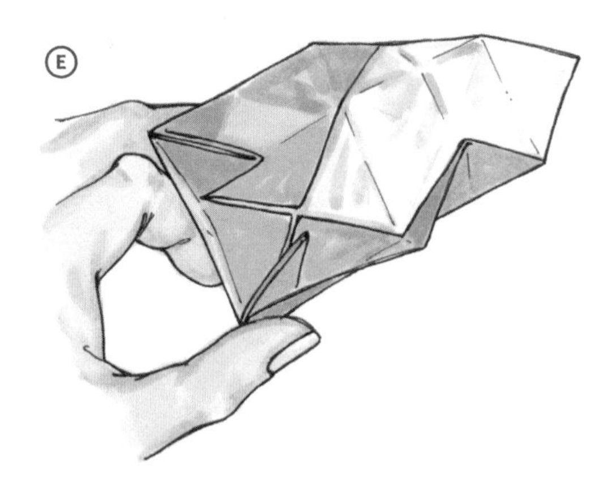

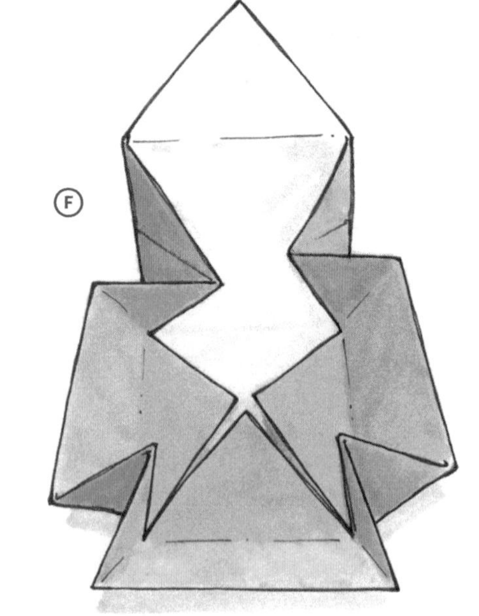

STEP 5: Repeat with the opposite side of the box (g), and you will have completed either the box or the lid (h).

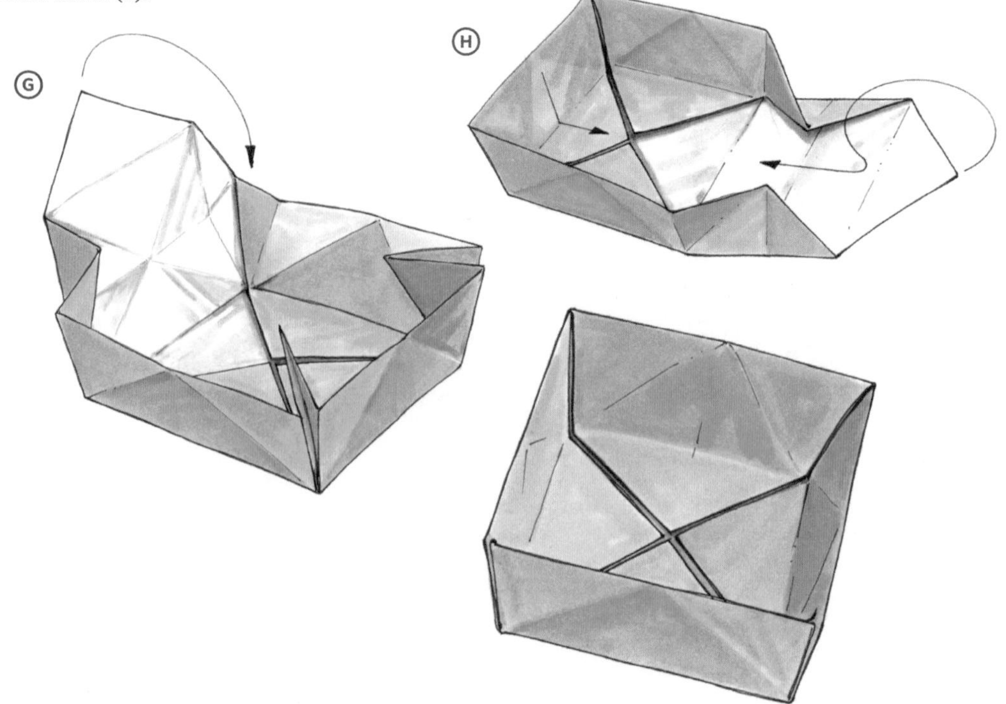

Make a Traditional and Thrifty Wreath

Cost savings

Between $12 and $35, depending on size and choice of greens

Benefits

A holiday wreath looks good and makes a nice gift or craft item

Create and customize a simple holiday wreath. You can purchase foam or wire wreath bases, but the *BackHome* staff shows you how to make your own for free here.

MATERIALS

- Rigid polystyrene foam (a piece of packing material is perfect)
- A seen-better-days T-shirt (preferably a dark color)
- Thumbtack or pushpin
- Craft knife or sharp knife
- Pencil
- String
- A big pail or basketful of holly or other greens, such as lignum vitae

STEPS

STEP 1: Create the wreath base.
Cut a ring from rigid foam or packing material (or foam insulation). Make the cut symmetrical by pushing a pin into the center of the block. Fasten this pin to a piece of 24-inch (61 cm) piece twine with a pencil tied to the other end. Trace around, making a circle. Next, shorten the twine by 4 inches (10 cm) and trace the inside circle. Cut away surplus foam with a knife.

STEP 2: Cover with fabric. Starting at the bottom hem of the T-shirt (fabric), cut a long strip about 1 to 2 inches (2.5 to 5 cm) wide, ending at the sleeves. Wrap this strip around the foam base as you would wrap a bandage. Firmly fasten the end with a thumbtack or pushpin.

STEP 3: Add greenery. Cut greenery into 4- to 5-inch (10 to 13 cm) lengths. Insert into slits between strips of material. Save smaller bits to fill in bare spots.

Bake Pumpkin Treats

Cost savings

Considerably less expensive than store-bought pasteries and breads, particularly if fresh pumpkin is used

Benefits

Fresh, healthy fare rich in beta-carotene

Thanksgiving dinner wouldn't be the same without dishes showcasing native squash: pumpkin. Here, *BackHome*'s cooking editor Judy Janes shares some sweet pumpkin recipes.

Pumpkin Fritters

Makes 12 fritters

- 1 15-oz (450 g) can pumpkin or equivalent (2 cups packed, fresh; see note)
- ½ cup (75 g) chopped sweet red pepper
- 1 egg yolk, well beaten
- 2 tablespoons (16 g) flour
- ½ teaspoon salt
- 1 egg white

Preheat the oven to 200°F (93°C, or gas mark ½). Lightly grease a large skillet and heat over medium heat. In a large bowl, mix together the pumpkin, red pepper, egg yolk, flour, and salt.

In a small bowl, beat the egg white until stiff but not dry. Fold the egg white into the pumpkin mixture and pour by rounded teaspoonfuls onto the skillet. Let cook for 2 to 3 minutes on one side, and flip over, cooking the second side for 2 minutes or so, until done. Transfer the fritters to a cookie sheet (not touching) and keep them warm in the oven until all of the fritters are done. Arrange on a heated platter and drizzle with honey or, even better, maple syrup.

Note: If using fresh pumpkin, be sure to cook down to the consistency of canned. Cut the shell into cubes and cook in the oven in a heatproof pan, for 20 minutes at 325°F (162°C) until tender. Mash with a fork to yield 2 cups packed pumpkin.

Pumpkin Bread

Makes 2 loaves

- 2½ cups (310 g) all-purpose flour
- 2 cups (250 g) whole wheat flour
- 2 cups (400 g) sugar
- 1 tablespoon (13 g) baking powder
- 2 teaspoons baking soda
- 4 large eggs
- 1 15-ounce (420 g) can pumpkin
- 2 teaspoons kosher salt
- 2 sticks unsalted butter, melted

Place the oven rack in the lower third of the oven and preheat the oven to 350°F (180°C, or gas mark 4). Grease two loaf pans and line them with parchment or waxed paper.

In a bowl, whisk together the flours, sugar, baking powder, and baking soda.

In a separate bowl, lightly beat the eggs, pumpkin, and salt. Add this mixture to the dry ingredients and mix until smooth. Mix in the butter a few tablespoons at a time.

Divide the batter between the two pans. Bake for 45 minutes to 1 hour, until the bread is golden and a wooden pick inserted into the center comes out clean. Invert it onto cake racks and cool completely. Store tightly covered in the refrigerator.

Pumpkin Bread Stuffing

This recipe should make enough to accompany a 14-pound (6 kg) turkey.

Makes 4 to 6 servings

- 6 cups (690 g) ½-inch (1 cm) cubes pumpkin bread, oven dried at 200°F (90°C, or gas mark ½)
- 6 slices bacon, cut up
- 1 cup (160 g) chopped onion
- 1 cup (100 g) chopped celery
- ¾ teaspoon salt
- ¼ teaspoon pepper
- ⅛ teaspoon dried sage
- 1 cup (235 ml) chicken broth, or more to taste

Preheat the oven to 350°F (180°C, or gas mark 4). Butter an ovenproof casserole dish.

Prepare the cubed pumpkin bread and set aside.

In a skillet, cook the bacon until crisp. Drain well on paper towels, reserving the drippings in the pan. In the same pan, sauté the onions and celery until soft. Add the pumpkin bread cubes, bacon, salt, pepper, and sage and mix well. Add the broth, a little at a time, until the stuffing sticks together. (You can form a tennis ball–size clump in your hands.) Place in a buttered ovenproof casserole. Bake for 50 minutes, until a thermometer reads 165°F to 170°F (74°C to 77°C).

Make the Season with Mulled Ciders

Cost savings

If made in bulk, costs can be half that of commercially prepared blends

Benefits

Healthier, fresher spice blends custom-made to your taste

Mulled cider is sure to warm the spirit during the holidays. Here, contributor Lynn Smythe offers a simple recipe.

Mulling Spice

- 1 gallon (3.8 L) apple cider
- 2 whole cinnamon sticks, broken into pieces
- 1 tablespoon (14 g) whole cloves
- 1 tablespoon (14 g) whole allspice berries
- Fresh orange peel from 1 orange
- Unsalted butter (optional)
- Ground nutmeg (optional)

Pour the apple cider into a large pot and place it over medium-high heat. Put the cinnamon, cloves, and allspice on a small square of cheesecloth and tie it closed with a piece of string. Place the spice bundle, along with orange peel, in the apple cider, and bring it to a boil. Immediately reduce the heat to low and simmer it at least 30 minutes before serving. To serve, if desired, top the mug of spiced cider with a small pat of butter and a sprinkle of ground nutmeg.

Can the Gingerbread Man

Cost savings

About 78 cents per cookie

Benefits

Homemade cookies decorated as you wish

There's nothing like an assortment of old-fashioned gingerbread man cookies to get friends and family members into the holiday spirit. These make great stocking stuffers, gifts, or decorations.

Large Gingerbread Man

Makes 24 large gingerbread men

- ¾ cup (165 g) butter
- ½ cup (115 g) brown sugar, packed firm
- 1 egg
- ¾ cup (255 g) molasses
- 3 cups (360 g) sifted flour
- ¼ teaspoon (1.5 g) salt
- ½ teaspoon (1.1 g) ground cloves
- ½ teaspoon (1.1 g) ground nutmeg
- 2 teaspoons (3.6 g) powdered ginger
- 1 teaspoon (2.3 g) ground cinnamon

Spray a cookie sheet with cooking oil. Beat the butter and brown sugar together; add the egg and molasses. In a separate bowl, combine the flour with the salt and spices and mix it into the moist ingredients. Chill for 2 hours. Roll out the dough and cut out cookies. Bake for 10 minutes at 375°F (190°C, or gas mark 5). Cool. Decorate with homemade icing.

Homemade Gingerbread Icing

- 1½ cups (180 g) confectioners' sugar
- 1 egg white
- Several drops lemon juice

Beat the ingredients until thick and smooth. The mixture should stiffen up like a paste; if it's too runny, add more confectioners' sugar. Add food coloring if you wish; mix well. Put it in a pastry bag and pipe onto cookies. Cinnamon red hots make suitable eyes and mouth; use icing as glue. Once the icing has hardened (1 hour or so), tie on a hanging ribbon (if using) and wrap each cookie securely in plastic food wrap.

Make a Seasoned Soup Gift for the Holidays

Cost savings

Each seasoning mix gift costs less than $3 to prepare

Benefits

An inexpensive homemade gift that is attractive and still very practical

Searching for the perfect gift—something that is unusual, inexpensive, and personal? Home comfort foods are the answer. Here, contributor Beth Lorms offers recipes.

A jar of ingredients for a nice, warm holiday soup is a perfect solution to the problem of small gift giving. Package them neatly—get creative and decorate lids with 7½-inch (19 cm) circles of fabric tied with ribbons or raffia. Finish with a tag that includes ingredients and baking instructions. You can also find decorative jar lids located in the canning section of your local food or hardware store.

Seasoned Bean Soup

- ½ cup (125 g) kidney beans
- ½ cup (112 grams) split yellow peas
- ½ cup (125 g) black beans
- ½ cup (96 g) red lentils
- ½ cup (125 g) small red beans
- ½ cup (112 g) split green peas

Layer each type of dried bean in a 24-ounce (113 g) jar. Mix together the Seasoning Mix (recipe at right) and place in a resealable plastic bag. Attach to the jar together with a recipe card for Seasoned Bean Soup.

Seasoning Mix

- 1 tablespoon (3.6 g) dried sweet pepper flakes*
- 2 teaspoons (4 g) chicken bouillon granules
- 2 teaspoons (5 g) dried minced onion
- 1½ (9 g) teaspoons salt
- 1 teaspoon (1.2 g) dried parsley flakes
- ½ teaspoon (1 g) black pepper
- ½ teaspoon (1.5 g) garlic powder
- ½ teaspoon (1 g) celery seed
- 4 tablespoons (60 g) brown sugar

*Dried sweet pepper flakes add a real a kick. If you prefer just a little kick, add fewer pepper flakes, or eliminate them entirely if don't want any kick at all.

Directions on card: You will need the following to prepare Seasoned Bean Soup:

Dried Bean Mix

- 2 14½-ounce (435 g) cans stewed tomatoes

Seasoning Mix (in bag)

- 1 teaspoon (5 ml) Liquid Smoke smoke flavor in a bottle (optional)

Rinse the beans and place in a large Dutch oven or stockpot. Pour 4 cups (950 ml) of boiling water over the beans, cover, and let soak overnight. Drain the beans and return to stockpot. Add 6 cups (1.4 L) of water, cover, and bring to a boil over high heat. Reduce heat to low, and simmer for 1 to 1½ hours, or until beans are almost tender. Add the tomatoes and Seasoning Mix. Stirring occasionally, cover and simmer for 30 minutes. Uncover the beans and continue to simmer for about 1 hour longer, or until the beans are tender and the soup thickens. Serve warm. Makes about 10 cups of soup.

Make Kids' Candleholders

Cost savings

Between $0.99 and $2.49, the cost of inexpensive candle holders

Benefits

Learning a craft and being creative and imaginative

Even the junior staff at *BackHome* magazine gets involved when it comes time to put together the holiday issue, and these simple and attractive candle holders are perfect for young hands. But don't take our word for it; let Zoë Anderson and Lily Quinn tell their story.

We made these decorative candleholders for Christmas presents, but together with friends or classmates, you can produce them as a moneymaking project for your school or organization.

MATERIALS

- 5-inch (2.5 cm) rounds, cut from sturdy material such as Masonite or other hardboard
- Hot glue gun
- Plastic bottle caps
- Red candles (of the same diameter as the bottle caps)
- Silver spray paint (optional)
- Walnuts and pecans in their shells
- Small pine cones
- Hickory nut hulls
- Dried holly berries and leaves
- Bittersweet branches
- Other found natural materials, such as small dried leaves or seed pods, and acorns

STEPS

STEP 1: Prepare bottle caps. Measure the bottle caps and make a note of their diameter. Use red candles of the same diameter. Find the center of each 5-inch (2.5 cm) round and mark it with an X in pencil. Glue a bottle cap over the X.

STEP 2: Add decoration. Glue other objects around the bottle cap, with larger objects like pinecones and walnuts close to the bottle cap and smaller items closer to the edge. Cover the surface with found objects.

STEP 3: Finish with paint. Place candleholders, one at a time, inside a tall cardboard box and spray pain according to directions on the can. (Be sure to do this in a well-ventilated area.) Allow candleholders to dry for 10 minutes. Finish by positioning the red candle in the painted holder.

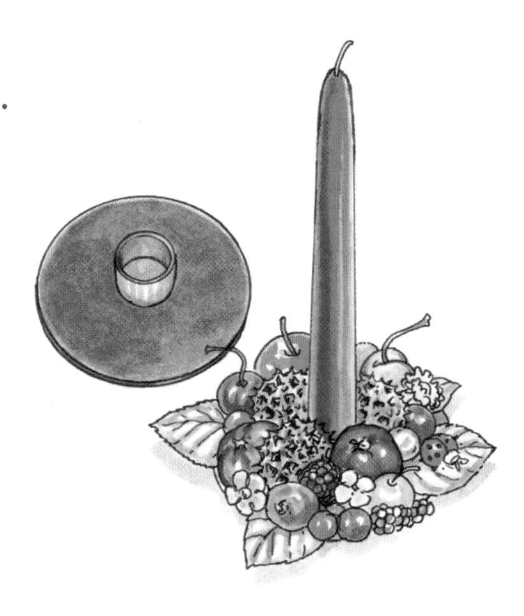

Mix Up "Bomb" Effervescent Bath Treats

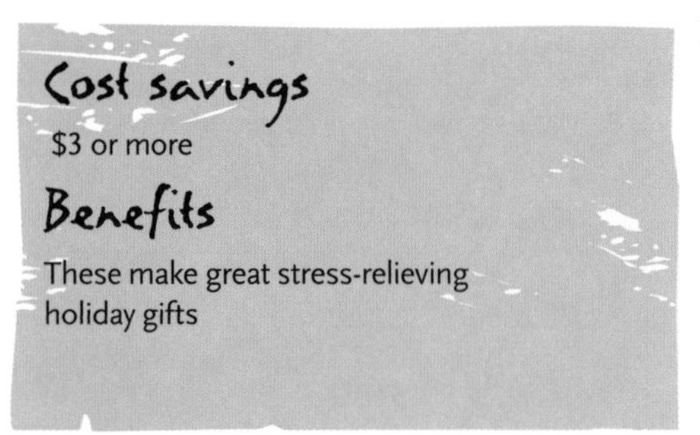

"Bombs" are easy-to-make, delightful-to-use gifts for the bath. Mary Ann Osby, whose husband, Don, is *BackHome*'s art director, offered this tutorial on how to make these clever little fizzers.

When it comes to a holiday gift project that's easy enough for children, yet one with enough creative freedom to challenge adults, you can't beat the bomb. This fizzy additive for the bath can be as simple or as elaborate as you like. You can blend bombs with scents and textures to please just about anyone on your holiday list. Plus, they're economical and fun to make.

Dropped into a full bathtub, a bomb will fizz and scent the water, and the baking soda will soothe away aches and fatigue.

Commercially, bombs are available for about $3 to $4 in bath and body-care shops. But once you discover how easy and inexpensive they are to make, you'll never buy them again.

To get started, you need a large mixing bowl and a fine-mist spritzing bottle. The recipe is simple. It's also flexible, so specific amounts aren't given. A good-size starting batch is 1 cup (235 ml) of citric acid (available from health-food stores) and 3 cups (660 g) of baking soda. You'll only need small amounts of coloring, flowers, and herbs, maybe 1/2 tablespoon (7 g) of each. For the oils, start with only around 3 to 6 drops.

Bath Bomb

- 1 part citric acid
- 3 parts baking soda
- Pinch of red clay or some food coloring (optional)
- Dried flowers or herbs (optional)
- Essential and fragrance oils
- Witch hazel

Cover several cookie sheets with waxed paper.

Place the citric acid and baking soda in a large bowl. Blend with a fork or your fingers until the mixture is lump-free. If desired, blend in the coloring and the dried flowers or herbs. Next add essential and fragrance oils drop by drop. Blend the mixture to achieve an even consistency.

Next comes the part where caution *must* be used. Rushing the process at this stage may set off the bomb prematurely. (You want to save the fizz for the bath, not the bowl.) Place the witch hazel in the spritzing bottle. As you blend, spritz the dry mixture two or three times and continue to knead. As it is worked in, repeat the process until you get a consistency that will hold together when squeezed in the palm of your hand. The process may take up to 10 minutes, but go slowly.

When the consistency is just right, you can start to form golf ball-size spheres or press the mixture firmly into small molds (such as small Jell-O molds). If molds are used you can check the mix by turning the mold over on a flat surface after about 5 or 10 minutes. If the bomb holds its shape, it's perfect. If it breaks apart, you may need to return the mix to the bowl and add a little more liquid.

After all the balls or shapes have been formed, place them on the cookie sheets and allow them to dry for about 12 hours. Then wrap the bombs in plastic or place them in an airtight jar decorated for the holidays.

Make Your Own Homemade Soaps

Cost savings

At about 98 cents per pound, homemade soap is around 8 percent the cost of commercial soap

Benefits

Healthy, nonanimal soap custom-made to your tastes and needs

Contributor Lynda McClanahan explains the benefits of making your own soap.

I first started making soap in the mid-1970s because I wanted to use only animal-free products. Over the years I've experimented with a number of variations, and occasionally, I'll invite a few friends over to make soap, and we all stir—and chew—the fat together.

ONE, TWO, IT'S NOT BLUE

There are two basic ingredients for vegetarian soap: one 13-ounce (369 g) can of granulated lye diluted in 2 1/2 pints (1.2 L) of water, and 6 pounds (2.7 kg) of vegetable oil. The lye is the kind used to unclog drains. It's best to purchase it new each time you make soap as old lye clumps and can be dangerous to work with. Lye (sodium hydroxide) in any form, however, is caustic must be treated with caution.

For the vegetable fat, I use olive oil, but almost any vegetable oil will do. However, peanut oil and solid shortening (such as Crisco) turn into something resembling frozen meringue. Coconut oil produces huge bubbles and rich lather (it's used in most commercial products), but it's drying to the skin. I use it sparingly, as it's also expensive. Whatever you use, since you won't be eating it, get the lowest grade possible.

I usually don't add scent or color to my soap, as I would rather have a clean product, and scent doesn't hold up well against the lye. If you must have color, use crayon shavings or burnt sugar. Some additions I *have* tried with success are castor oil (it produces a soft soap with excellent healing properties), chlorophyll capsules (one or two cut and emptied give a nice blue-cheese effect and cut down on the "bacony" homemade smell), glycerin (just a bit), and vitamin E (only a few capsules). Over the years I've added powdered milk, tea, oats—you name it. (A bit of oats added just before molding produces a pumicelike soap.) If you experiment, make sure the ratio of oil to lye to water in the recipe remains the same.

THREE, FOUR, A FEW THINGS MORE

The utensils you need are items you probably already have or can obtain easily:

1. A kitchen scale marked in ounces.
2. A large enamel, glass, or ceramic pot (not aluminum—you'll ruin it). I use an enameled canning vessel.
3. A good-sized heat-tempered glass container. Large apple juice jugs work well, since the jars are sterilized with heat at the factory. Clean the jar thoroughly and poke two holes in the lid with a 16-penny nail.
4. A wide-mouthed funnel for the lye. I use the top-with-handle section cut from a plastic bleach bottle.
5. A long-handled wooden spoon.
6. Rubber gloves, safety goggles, and a sturdy apron. Lye is extremely caustic, so wear the gloves throughout the process, and work slowly to avoid splashing and drips.

7. A waterproof thermometer.
8. Molds. I use milk cartons with the opened end stapled shut (or the cap on) and two big squares cut out of one long side. This leaves a strip of carton holding the mold together, yet allows air circulation so the soap can cure. These cartons are easy to work with, since the waxed coating helps the paper carton tear off cleanly. You can also try shoe boxes lined with plastic. Whatever you use, assemble it before you start the soap. If molds are not ready, you'll end up with pounds of raw soap setting up in the pot with no place to put it.

FIVE, SIX, TIME TO MIX

First clear your sink and countertop. Cover the area with plastic shower curtains or tablecloths (which you can wash off, wearing gloves, when you're done) or trash bags or newspaper, which can be thrown away.

Put the 2 1/2 pints (1.2 L) of water into the glass container; using the funnel, ***slowly and carefully*** pour the granulated lye into the water. (Lye is not pleasant to work with and has a tendency to hang in the air, where it's hard not to breathe it in.) Stir with the long handle of the wooden spoon until all the granules disappear. If the lye has become damp, it may take a lot of stirring to dissolve completely. Once the lye is added to water, the solution becomes quite hot (about 200°F, or 93°C); place the jar in your sink, screw on the vented lid, and proceed to the next step. At this stage, especially, lye will burn a hole in anything—clothes, skin, flooring, etc.

Weigh your pot, then add oil until the scales read exactly 6 pounds (2.7 kg) plus the weight of the pot. Heat the oil in the pot to 97° to 99°F (36° to 37°C). Since oil tends to heat faster than lye cools, the trick is to get both the lye and the fat to reach the same temperature in order to combine them. I usually start the lye pretty far ahead of the oil, but either container, or both of them, can be cooled in cold water (a double sink is handy for this).

When the elements are both at the correct temperature, carefully pour the lye solution—through a hole in the jar lid—into the oil, stirring in an unhurried but steady fashion. You should notice the oil getting slightly less clear, then slowly turning a creamy color.

Now stir. Soap can take from 20 minutes to an hour to set up. There's no need to beat the stuff to death; you're just trying to get as much fat to come into contact with as much lye as possible. Slow and steady wins the race, and don't forget to make regular swipes through the center of the pot. Keep at it until the soap has the consistency of runny pea soup. Use a ladle to drizzle a design onto the bowl of the spoon. If the design holds for a few moments, the soap is ready to mold. As a rule, I don't stir for more than an hour. If the soap isn't ready by then, I figure I've done something wrong.

SEVEN, EIGHT, POUR AND WAIT

Although you've got soap at this stage, it's uncured and green; it'll still burn like the dickens if you get some on you. So, again, be careful as you fill your molds. Set them in a reasonably warm place for 24 hours. Folk wisdom decrees insulating the soap with blankets for this period, but I've found that most modern houses are warm enough to keep the soap from cooling too quickly, which is the point of the warmth.

After a day or so, check whether the soap is hard enough to be removed from the molds. Since curing requires air circulation, the sooner you do this, the better. Leaving soap in the molds won't hurt it; it'll just slow the curing process.

Once you've torn the molds off the hardened blocks, you can cut the soap into bars. At this fairly soft stage, leftovers can be pressed into little balls that make nice gifts. But wait at least three weeks before using the soap.

NINE, TEN, THAT'S THE END

In a pinch you can utilize your soap for dishes and laundry, though commercial products tend to spoil you for this. Simply dissolve 1 pound (450 g) of your soap shavings in 1 gallon (3.8 L) of water, and heat into a soft, jellied consistency. By diluting further you can also make a good spray to use on pesky aphids in the garden. When old-timers threw dishwater on their roses and fruit trees, they weren't using modern detergents. And since the price of commercial insecticidal soap is incredibly inflated, you can save a lot of money, especially if you have a good-sized orchard.

When Things Go Wrong

1. If the soap doesn't set up in the pot, don't despair. As long as you've measured everything correctly, you should be able to salvage the situation. Maybe the mixture is too cold. Gently heat it and try again. If that doesn't work, try pouring it into the molds and leave it for a while. It still may harden in time.
2. If the soap turns creamy but doesn't pass the design test, keep stirring or pour into the molds. You can always throw a batch out, but you may as well try to cure it first.
3. Sometimes clear, caustic, runny stuff sloshes out of air bubbles and corners when the soap is removed from the mold. Relax: just wash your hands and pour it off. This is simply water and lye that didn't turn into soap.
4. It's normal for a fine, powdery ash to form on the surface of the hardened soap. This is no big deal, but it's extremely drying to the skin, so scrape it off before using the soap.
5. Homemade soap is softer than commercial products, even after curing. This is also normal, as factory-made soap is usually milled or made under pressure. Your soap isn't, and neither was your grandmother's. Enjoy it.

Make Old-Fashioned Jellies

Cost savings

About $8, or the cost of a jar of purchased boutique jelly

Benefits

Homemade jellies make wonderful gifts.

Contributor Jane Scherer has put together recipes for four delicious jelly samplers that make delightful holiday gifts.

These jellies are not your usual fare, but the ingredients are common. All you need is sterilized jelly jars with lids that seal under heat, and a jelly bag or strainer for the parsley jelly recipe.

Parsley Jelly

This old English recipe is very good with meats. I prefer using Italian parsley, if possible.

Makes about 5 half-pint (115 g) jars

- 4 cups (240 g) washed and chopped parsley
- 3 cups (705 ml) boiling water
- 2 tablespoons (28 ml) lemon juice
- 1¾ ounces (50 g) powdered fruit pectin
- 4½ cups (900 g) granulated sugar
- A few drops green food coloring

Place the parsley in a bowl. Add the boiling water, cover, and let it steep for 15 minutes. Pour into a jelly bag and squeeze out the liquid. Or, pour the liquid through a strainer, reserving the liquid, and press the parsley with a large spoon. Measure 3 cups (705 ml) of the liquid into large saucepan. Add the lemon juice and pectin and mix well. Place over high heat, and stir until mixture comes to a hard boil. Stir in the sugar. Add enough food coloring to tint lightly. Bring to a full rolling boil and boil for 1 minute, stirring constantly.

Remove from the heat. Skim off the foam with a metal spoon and pour quickly into sterilized jelly glasses. A hot water bath is recommended for safety: Set filled and lidded jars into a large pot with a rack and boil for 5 minutes. Remove the jars, let them cool, and tighten the lids.

Garlic Jelly

This is good with roasts and chicken and on crackers, and it's also good for you.

Makes about 5 half-pint (115 g) jars

- ¼ cup (58 g) peeled garlic cloves
- 2 cups (470 ml) distilled white vinegar
- 5 cups (1 kg) granulated sugar
- 3 ounces (90 ml) liquid pectin

In a food processor or blender, blend the garlic and ½ cup (120 ml) of the vinegar together until smooth.

In a large pot, combine the garlic mixture, 1½ cups (350 ml) of the vinegar, and the sugar. Bring to a boil, stirring constantly. Add the pectin and return to a boil. Boil hard for 1 minute, stirring constantly. Remove from the heat. Fill sterilized jars with the jelly, and process as described.

Rose Geranium Jelly

Geranium leaves are free for the picking, though you should be certain they haven't been treated with pesticides. The jelly is pleasantly fragrant and has a delicate taste.

Makes about 10 half-pint (115 g) jars

- 1 quart (946 ml) unsweetened apple juice
- 10 cups (2 kg) granulated sugar
- Juice of 1 lemon
- One 6-ounce (185 ml) bottle or two 3-ounce (85 g) packets Certo (natural fruit pectin)
- 15 rose geranium leaves (fresh or frozen)

Put the apple juice, sugar, and lemon juice in large pot. Bring to a full rolling boil. Add the pectin and boil again. Remove from the heat. Add the rose geranium leaves. Leave in the hot liquid for 1 minute. Remove the leaves with a slotted spoon and ladle the jelly into sterilized jars. Process as described.

Craft a Corncob Doll from Scraps

Cost savings

The doll is made from scrap and recycled material and costs nothing

Benefits

A unique, handcrafted toy

In the pioneer days, toys were simple and mostly handmade. Doll dishes were crafted from clay, doll-sized clothing and furniture were made from fabric and paper scraps, and entertainment was where you found it. Here, *BackHome* contributor Ann Heatherley describes an old-time doll-crafting technique from her grandmother's childhood.

Clara Ann Morris was my grandmother. Around 1870, her parents moved by covered wagon from Tennessee to Texas, and three years later she was born in a log cabin shared with three older brothers and two younger sisters. With so many mouths to feed, everything was done with a purpose and nothing was wasted.

She told me many stories of her life as a pioneer girl, and she made these images come alive for me. Once, when she and my mother were making quilts, she taught me to make a doll-size one. Best of all, though, was the corncob doll my grandmother made for me, dressed in clothes made from scraps. This doll was more special than any other I had because it was just like one my grandmother had played with, and it was clothed in the familiar fabric of one of my grandmother's old dresses.

In making this doll, you may find yourself passing along a bit of yesterday to be enjoyed tomorrow. One note: Though it wouldn't be a corncob doll without a cob, this doll can be made with any item of generally the same size and shape, such as a piece of wood or a tight roll of fabric.

MATERIALS

- 1 corncob (dried)
- 1 skin-colored sock
- Scraps of material for clothes
- Embroidery thread, paint, or crayon for the face
- Needle, thread, and scissors

STEPS

STEP 1: Create the doll body. Lay the cob on the foot portion of the sock, and measure the amount of sock needed to cover the cob. Allowing for extra on the raw edges to tuck under, cut to fit snugly. Wrap the sock around the cob, turn under the raw edges, and whipstitch in place (a). The seam will become the back of the doll.

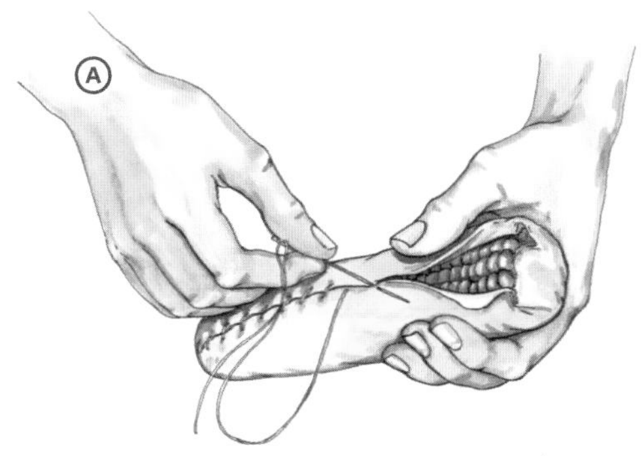

STEP 2: Make doll arms. Cut two ½ x 1½-inch (1.3 x 3.8 cm) strips from the remaining sock. Stitch one end and the side of each piece. Turn the pieces right side out, and stuff them with a bit of cotton or scraps of sock. At the open ends, turn the raw edges under, and sew an arm onto each side of the cob, 1 ½ inches (3.8 cm) below the top. With strong thread, wrap several times around each arm ½ inch (1.3 cm) from the end to form a hand (b).

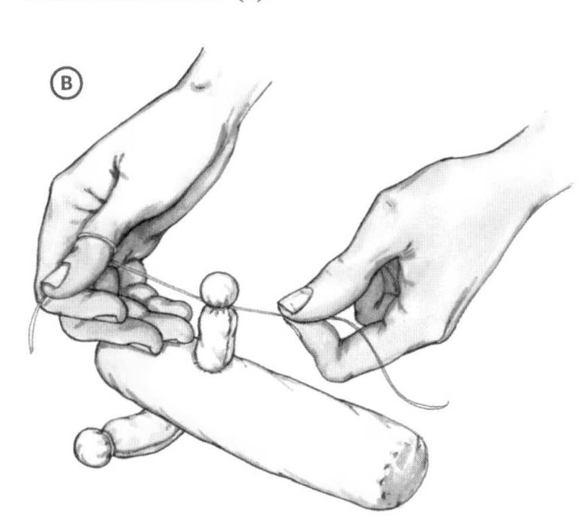

STEP 3: Make the face. You can embroider (c) or use paint or crayon. Prevent crayon from smearing by laying a paper towel over the crayoned face and pressing it with a hot iron for a few seconds.

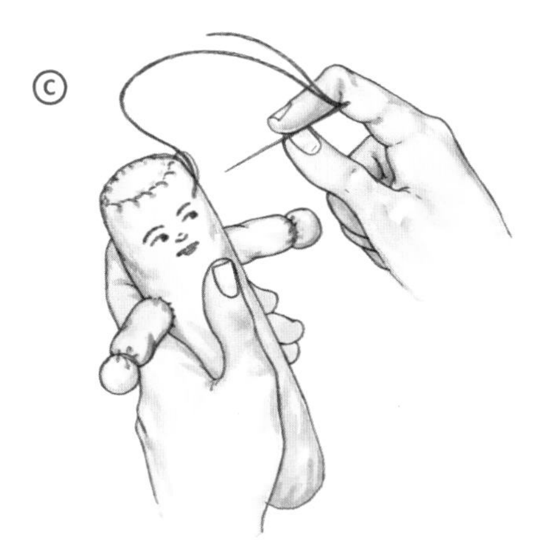

STEP 4: Make the clothes. Use the patterns here, making adjustments to fit your corncob. Most pieces are sewn together wrong side out. Keep a hot iron handy to press the pieces after they are turned right side out.

SEW DOLL CLOTHING

Follow these instructions to make a dress, apron, and bonnet for your corncob doll.

Dress bodice

Using a piece of material at least 3 x 4 inches (7.6 x 10.2 cm), fold it in half lengthwise, then fold that in half crosswise. Lay the pattern on the folds as indicated, and cut the material. With the right sides of the material facing each other, sew the sides and underarms. Cut the fold down the back. Turn the neck under, and stitch. Turn the edges of the armholes under, stitch, and gather to fit the arms.

CUTTING AND SEWING DIAGRAM

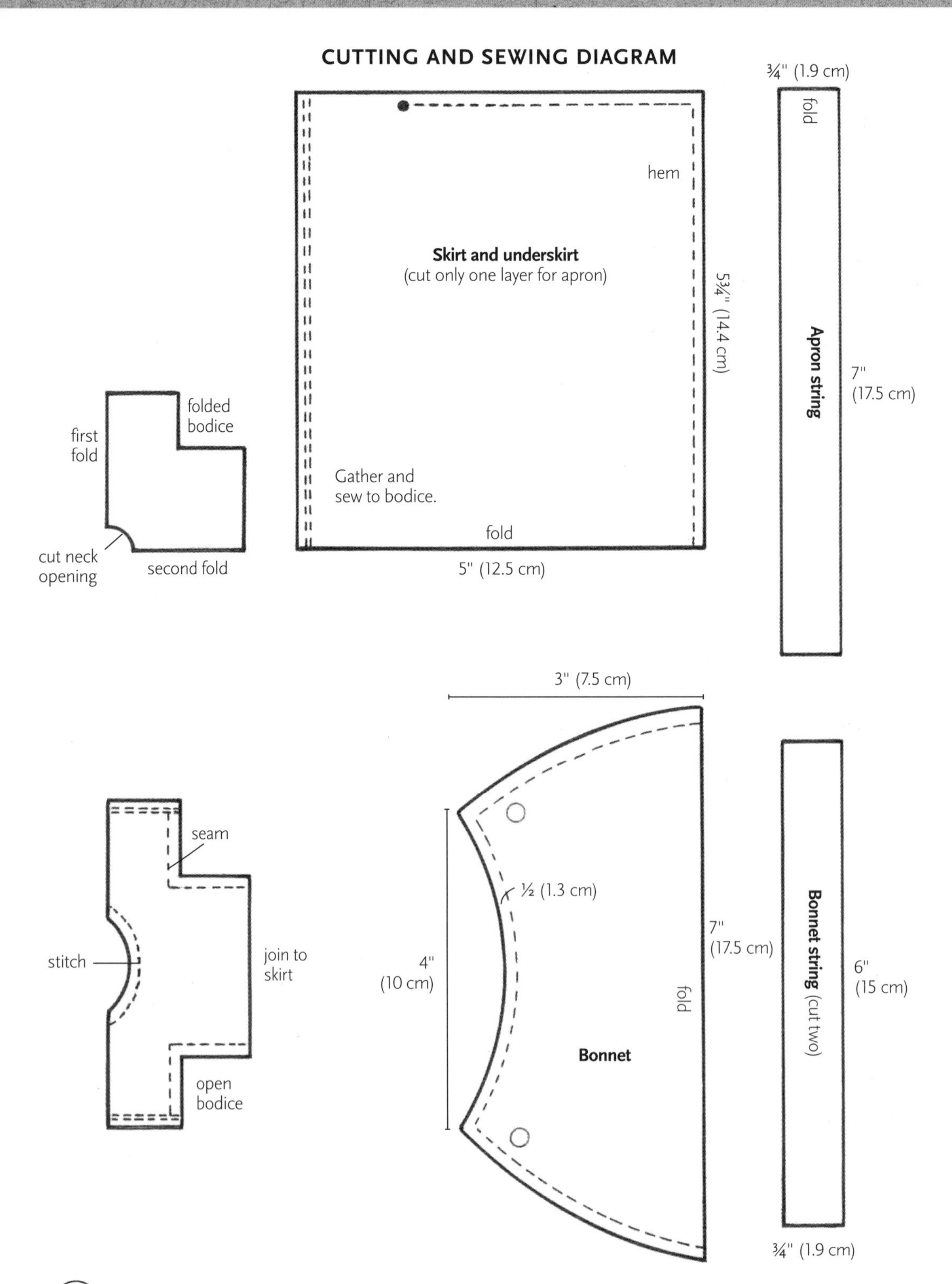

Dress skirt

Use a piece of dress material about 6 x 12 inches (15.2 x 30.5 cm). Fold it in half, lay the pattern on the fold, and cut the material. Turn the hem under, and stitch. Gather the skirt top until it fits the lower edge of the bodice, but leave $^{1}/_{4}$ inch (6 mm) ungathered on each side. With right sides together, sew the skirt to the bodice, with the open seams of the skirt matching the open back of the bodice. Sew the lower back seams of the skirt together (right sides facing), to the point indicated by the black dot on the pattern. Then turn under the raw edges of the bodice and open part of the skirt, and stitch each side separately. The dress can now be put on the doll. Either stitch it in place or add snaps at the neck and waist.

Dress underskirt

Make this the same as the skirt, except turn the top edge under before gathering it. It, too, can be stitched closed or snapped.

Apron

This is half the size of the skirt, so only one layer is cut from the skirt pattern. Turn under the raw sides and bottom, and stitch. Gather the top to fit just the front of the doll. Cut a $^{3}/_{4}$ x 14-inch (1.9 x 35.6 cm) piece of material for the apron strings. Center the apron on the strip, right sides together, and sew a seam where the apron meets the strip. Fold the strip lengthwise, right sides facing, and sew the seams on either side of the apron as well as across the ends. Turn the strings right side out, and slip-stitch the inside turned edge over the apron's raw edge.

Bonnet

You'll need a piece of material about 4 inches (10.2 cm) square for the bonnet and strings. Fold the piece in half, lay the bonnet pattern on the fold, and cut. Mark on the material the locations of the black dots and open dots. With right sides facing, stitch along the seam line, leaving an opening large enough for turning. Turn right side out, and slip-stitch the open edge. Bring the two outside black dots to the center dot, barely lap one side over the other, and stitch through all layers. For bonnet strings, cut two $^{3}/_{4}$ x 6-inch (1.9 x 15.2 cm) pieces of material. Sew seams, either with right sides facing and turning, or with wrong sides together, folding raw edges under. Attach a string to each side of the bonnet at the open dot.

PART 3

HEADING OUTSIDE: THE GREAT OUTDOORS

CHAPTER 9

ENTRYWAYS AND EXITS

A home's entryways and exits leave a first and lasting impression on visitors. Keep these fringe areas tidy and warm so guests feel welcomed. This chapter will help you keep entries and exits organized and insulated.

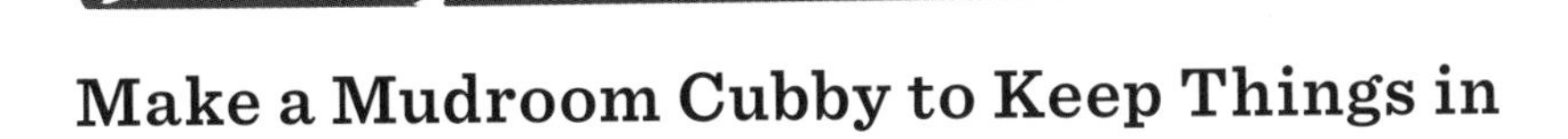

Make a Mudroom Cubby to Keep Things in Their Place

Cost savings

$70 or more in not buying other storage units or cabinets

Benefits

Organization of clutter, cleaner household floors, extends the life of your flooring, and gives a more relaxed feeling to your home

Build this simple shoe cubby to neatly store shoes. Contributed by Kellie Janes, this cubby is easy to make, and compartment size can be altered to fit shoes and boots of all sizes.

MATERIALS

- 3/4-inch (1.9 cm) shelving or wide boards long enough to yield two 29 3/4-inch (75.5 cm) lengths of wood and two 25-inch (63.5 cm) lengths. If using 3/4-inch (1.9 cm) plywood, cut the widths to 12 inch (30.5 cm). Dimensional 1 x 12 boards from the lumberyard are 11 1/4 inches (28.1 cm) in width.
- Two 8-foot (2.4 m) lengths of hardboard or finish plywood in 12-inch (30.5 cm) width
- One 3-foot (90 cm) lengths of 1/4-inch hardboard or finish plywood in 12-inch (30.5 cm) width (Note: You could cut all of these from a single 4 x 8 foot [1.2 x 2.4 m] panel.)
- Sander or drawknife (to smooth edges)
- 1 1/2-inch (3.8 cm) brads or finishing nails
- Carpenter's glue
- Jigsaw or table saw

STEPS

STEP 1: Sand edges. Take the four 3/4-inch (1.9 cm) pieces, first round one long edge of each with a sander or drawknife. This need not be a fancy job; you just want to take off the sharp edge. These pieces will become the cabinet frame and will be the most visible.

STEP 2: Join pieces. Join the 3/4-inch (1.9 cm) pieces by setting one of the short pieces on top of the two longer ones. Align the corners and edges. Drive 1 1/2-inch (3.8 cm) brads or finishing nails into the sides from the horizontal surfaces. (Use carpenter's glue to strengthen joints prior to nailing.) Flip the case over and repeat these steps on the bottom side.

STEP 3: Make cubbyholes. Cut the hardboard into four 29 3/4-inch (75.5 cm) and four 23 1/2-inch (59.7 cm) lengths. (The four

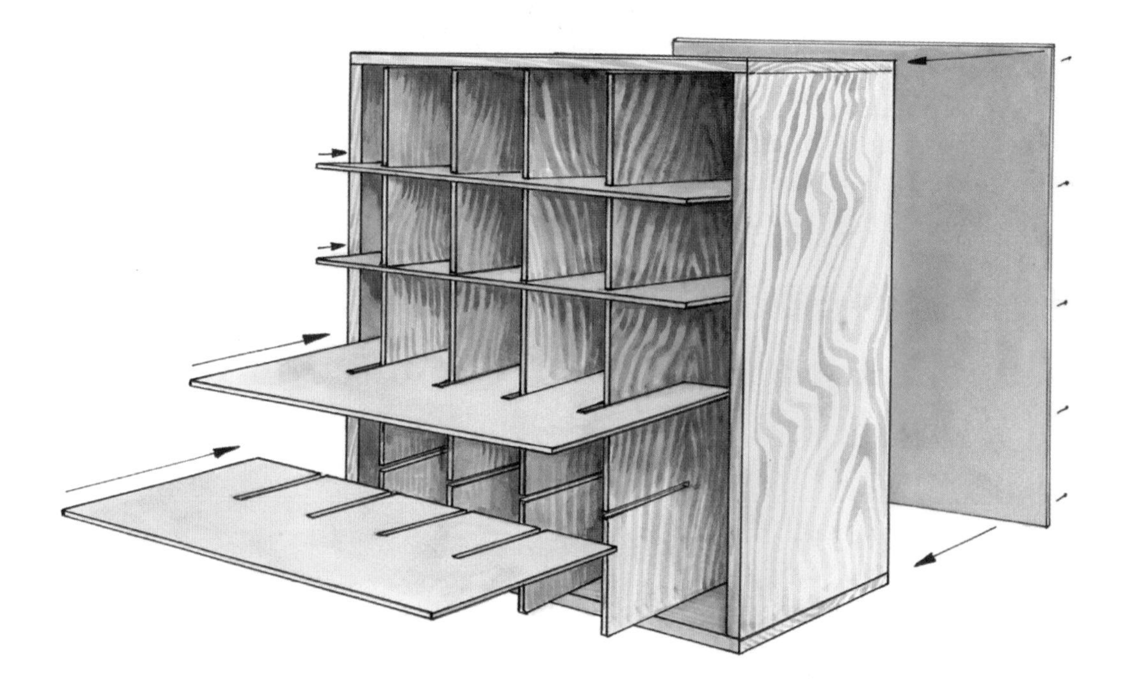

shorter pieces can come from one 8-foot [2.4 m] length, as will three of the longer ones.) Lay out the four 29 3/4-inch (75.5 cm) panels on a flat surface. Measure 5 3/4 inches (14.6 cm) from one end of each, and mark with a pencil. Then measure 1/4 inch (6 mm) and make a mark. Repeat this long-short pattern until the board is completely marked.

Next, lay out the four 23 1/2-inch (59.7 cm) boards and follow the same procedure, except this time measure 4 1/2-inch (11.4 cm) and 1/4 inch (6 mm). Finally, mark a centerline at each of the 1/4-inch (6 mm) marks for reference.

With that done, use a jigsaw or a table saw to cut out the 1/4-inch (6 mm) slots, just to the center mark. Sand the edges and slide the horizontal shelves into the vertical uprights, then fit the whole assembly into the cabinet frame.

STEP 4: Assemble optional hardboard back. You can leave the egg-crate insert unfastened for easy removal and cleaning, or you can cut a 24 x 31-inch (60.9 x 78.7 cm) piece of hardboard from the leftover scrap and tack it to the back of the cabinet. Secure it with glue.

Build an Entry Vestibule to Ease Your Winter Budget

Cost savings

$50 to several hundred dollars per heating season, depending on climate

Benefits

Fewer winter drafts, heating season savings, removable in warm seasons

How do you keep cold out of the biggest hole in the house? You can't seal your door shut, and while storm doors prevent drafts, they don't prevent cold air from entering when the door is open. One solution is an old-fashioned vestibule: a sort-of enclosed outside atrium that serves as an intermediary between the cold outdoors and warm living area. If you have an enclosed porch or patio or a covered breezeway, you already have a vestibule on a grand scale. The one described here is a lot simpler, costs a lot less, and can be dismantled and stored away in the spring when it's not needed.

Start to make this quick and inexpensive entry buffer by gathering some materials (a lot of this can be salvaged or "repurposed" from second-hand stores).

MATERIALS

- Three 36-inch (91 cm) surface-mounted aluminum-panel storm doors with hardware and glass intact
- Eleven pressure-treated 2 x 4s
- One 4-foot (1.2 m) piece of 1 x 4
- One 10-foot (3 m) piece of corrugated roofing
- A section of metal ridge cap to match the roofing
- 1-inch (2.5 cm) Styrene construction foamboard (optional)
- Level
- No. 8 x 3-inch (7.5 cm) decking screws
- 1/2-inch (1.3 cm) anchor shields
- Reclaimed aluminum entry threshold
- 1/8 x 1 x 1 inch (0.3 x 2.5 x 2.5 cm) aluminum angle
- 3/4-inch (6 mm) sheet screws
- 16 "penny sinkers" or No. 8 x 2 1/2-inch (6.5 cm) screws
- 1 3/4-inch (4.5 cm) gasketed roofing screws
- No 6. x 3/4-inch (1.9 cm) panhead screws
- Caulk
- Circular saw with a masonry blade (or a reciprocating saw with a metal-cutting blade)
- Canned foam sealant

STEPS

STEP 1: Create a frame. Measure the perimeter of the storm door frames. Cut the 2 x 4s to match the sides, top, and thresholds of two doors. Nail the lumber in stud-and-plate "box" fashion, with the top and bottom pieces resting on the ends of the side pieces.

Then, set the two open frames against the right- and left-hand casings of the entry door, plumb them with a level, and mark their position. Square them to the wall and mark position on the porch deck or patio at the bottom. Fasten the pieces at the door casing with No. 8 x 3-inch (7.5 cm) decking screws. Screw bottom plates to the deck or attach to concrete using $1/2$-inch (1.3 cm) anchor shields. Cut two more 2 x 4s to length and turn them on their edge. Fasten them to the inside of each frame unit at the bottom to add support.

To support the frames at the top, bridge the opening with a length of 2 x 4 cut to keep the vestibule walls parallel. At the bottom, you can use a reclaimed aluminum entry threshold to span the frames, fastening them to the deck with two pieces of $1/8$ x 1 x 1-inch (0.3 x 2.5 x 2.5 cm) aluminum angle and $3/4$-inch (6 mm) sheet metal screws.

STEP 2: Support roofing material. Three sets of simple rafters will support the lightweight roofing material. Rafters are positioned at the front, center, and rear of the structure. Cut rafters from 2 x 4s to 26-inch (66 cm) lengths, marking the angles at the ends so they can be joined flush to a 40-inch (101 cm) length of 1 x 4 ridge board. (That angle will depend on the pitch of the small roof you are fashioning.) Attach the tail ends of the six rafters to the tops of the wall frames with 16-penny sinkers or No. 8 x 2 1/2-inch (6.5 cm) screws, driven at an angle.

STEP 3: Construct roofing. Use a circular saw with a masonry blade (or a reciprocating saw with a metal-cutting blade) to cut four 26-inch (66 cm) pieces of the corrugated metal roofing. Also cut a 40-inch (101 cm) section of ridge cap. Butt the first two roof panels against the house and fasten them to the first rafter set with 1 3/4-inch (4.5 cm) gasketed roofing screws. Then overlap the second pair of roof panels and fasten them to the remaining rafter sets. Screw the cap over the ridge opening at the six rafter points.

STEP 4: Insulate the structure. Run a bead of exterior caulk along the back of each storm-door frame, then fasten the aluminum panels to the wooden frames using No. 6 x 3/4-inch (1.9 cm) panhead screws. At the front, fit the door so it sweeps the threshold evenly and secure it in place. Fill the triangular gable opening over the door with a section of scrap 1/2-inch (1.3 cm) or 3/4-inch (1.9 cm) plywood cut to fit.

Fill spaces between rafters with cut plywood scraps and seal with canned foam sealant. This expanding foam can also be used to seal any gaps and openings in the frame, or spaces where it meets the house.

If you want to insulate the roof, however, cut some pieces of 1-inch (2.5 cm) Styrene construction foamboard to size and fasten them to the rafters from the inside.

Stop Dirt in Its Tracks with a Simple Slatted Door Mat

Cost savings

Between $31 and $106

Benefits

A home-made dirt-catching entry or mudroom mat that snags debris

You can make a really functional door mat from scraps of wood (that's right, wood) that will actually do something those fancy mats cannot—and for just a few bucks. You may have seen variations of these slatted mats around, and at least one company makes them out of cypress (cost: more than $100).

The nice thing about a slatted configuration is that dirt, debris, and clumps of soil fall between the slats, where they remain out of the way until you lift the mat and sweep out from under it. These mats are ideal for a mudroom.

MATERIALS

- 15 strips of wood 3/4-inch (1.9 cm) wide, 1 inch (2.5 cm) thick, and 28 inches (71.1 cm) long
- Table saw
- 1/4-inch (6 mm) (inside diameter) neoprene hose or fuel line
- Utility knife
- Leather gloves (for protection)
- Three lengths of 3/16 x 24-inch (0.5 x 61 cm) threaded rod
- Six 3/16 x 24-inch (0.5 x 61 cm) nylon-insert stop nuts
- Hacksaw
- Linseed oil finish (or deck stain)

STEPS

STEP 1: CUT wood slats. You can use natural woods such as redwood, cedar, and cypress, and those must be cut to size. If you use treated one-by boards, these are already 3/4-inch (1.9 cm) wide, so all you need is a single 8-foot (2.4 m) 1 x 4 from the lumber yard. Cut slats that are 3/4 inch (1.9 cm) wide, 1 inch (2.5 cm) thick, and 28 inches (71.1 cm) long.

STEP 2: Mark and drill points. Measure and mark one point on the 1-inch (2.5 cm) face at its center point—14 inches (35.6 cm). Mark points 2 inches (5 cm) in from each end. Then use a 7/32-inch (6 mm) drill bit in an electric drill to bore through the wood at those three points. After the holes are made, you can use the drilled piece as a template to bore through the other 14 slats at the same locations.

STEP 3: Create "thread." Use a sharp utility knife (and protect hands with gloves) to cut a section of 1/4-inch (6 mm) (inside diameter) neoprene hose or fuel line. Cut 42 sections, each 1/2 inch (1.3 cm) in length, from the hose.

Take your three lengths of 3/16-inch x 24" (0.5 cm x 61 cm) threaded rod and six 3/16 x 24-inch (0.5 x 61 cm) nylon-insert stop nuts. Thread one nut onto the end of each of the rods so that the rod ends are flush with the ends of the nuts.

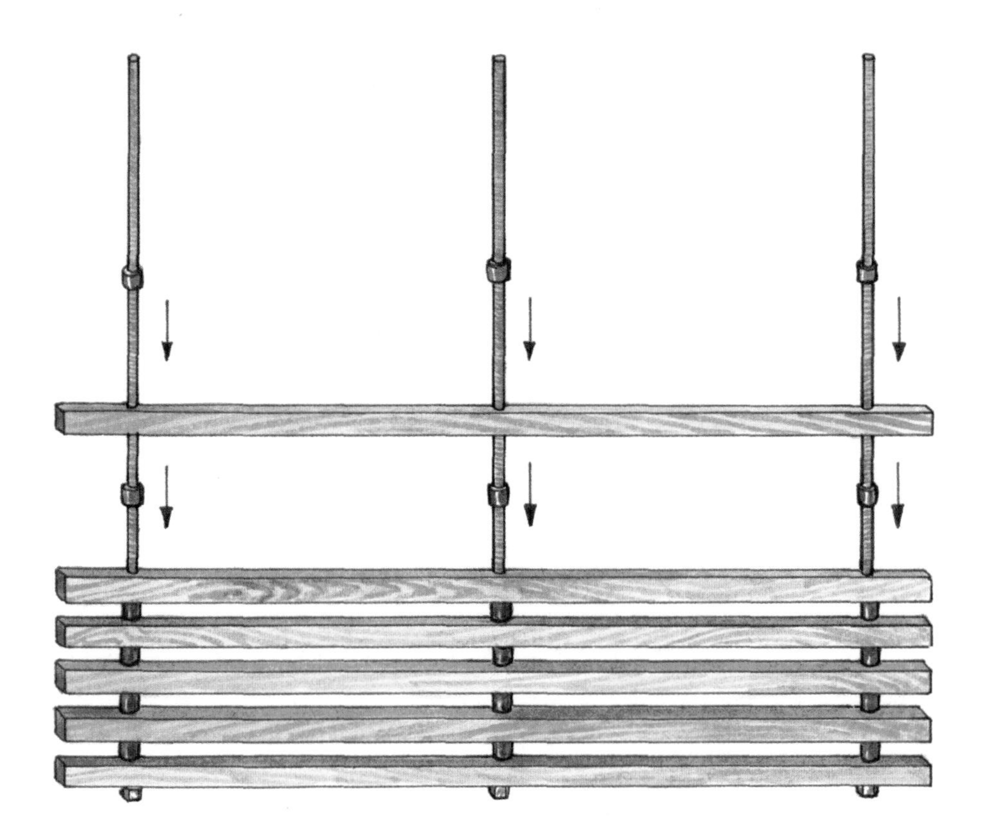

STEP 4: Thread the slats. Slip the rods through the holes in one of the wood strips until the strip rests against the edge of the nuts. Slide a rubber section over each of the rods. Then slip another strip over the rods. Continue in this alternating pattern until the last strip is in place.

Thread the remaining nuts onto the stubs of the three rods until they draw up snugly against the last strip of wood. (If the stop nuts are difficult to thread, lubricate the rod with silicone spray and use needle-nose pliers to grip the rod at a point between the slats; do not grip the rod at a point where the nut has to pass over it; you'll likely damage the threads.)

STEP 5: Finish the mat. Once the nuts are snugged, use a hacksaw to cut off the excess rod, close to the nut. You can put a linseed oil finish on the wood to help preserve it, or brush on some deck stain if you want to add some color.

CHAPTER 10

GARAGE AND WORKSHOP

Having a place to putter is part of the appeal of home, and having the skills to paint, decorate, and make simple repairs without calling in a professional is encouraging and cost-saving. In this chapter, we'll discuss at tools and supplies that will keep your home-repair budget in check while allowing you to keep up with the Joneses.

Find the Ten Must-Have Tools Cheaply

Cost savings

Between $50 and $90 if you buy at yard sales or flea markets

Benefits

Owning a decent set of useful tools

If your go-to tool is a butter knife, it's time to upgrade. Having the right tools makes household repairs a lot easier. Here, contributor Merwyn Price tells you how to find decent tools at most discount stores, hardware stores, home centers, and even flea markets and auctions. Expect to spend between $50 and $100 for a modest set of tools. You'll usually be safe buying recognizable name brands in the medium price range.

Screwdrivers: Avoid the temptation to buy the cheap plastic 99-cents-per-dozen kind that bend while tightening a pot handle. A ten- or twelve-piece set is nice, but the following four screwdrivers are probably the most useful around the house: A 3/16 x 6-inch (0.5 x 15 cm) cabinet (thin slotted); a 1/4 x 8-inch standard (thick slotted); a No. 1 x 4-inch (10 cm) Phillips; and a No. 2 x 4-inch (10 cm) Phillips. Screwdrivers with traditional cylindrical fluted handles do well because the shape is consistent.

Cost: $10 to $18 for the set

Hammer: Select a 16-ounce (455 g) claw hammer (the kind with the curved back for pulling nails) with a fiberglass or steel handle. A rip hammer has a straight cylindrical head, but the claw hammer is better for general use. A 10-ounce (280 g) hammer is too light for general purposes, and a 20-ounce (570 g) model has more heft than you really need. A wooden handle is nice but requires some care and is susceptible to splitting and gouges.

Cost: $14 to $17

Measuring tape: Don't bother with a small gimpy one. Get a 1-inch (2.5 cm)-wide blade that measures 25 or 30 feet (7 to 9 m) overall. Why? You'll understand the first time you measure a room for carpet or picture window for blinds. The 1-inch (2.5 cm)-wide blade allows one-handed measuring, and the aggravation of using a thinner, more flexible ruler blade isn't worth the savings of a dollar or two.

Cost: $8 to $11

Adjustable wrench: Crescent is a popular brand, and the generic name for a wrench of this type. An 8 or 10-inch (20 to 25 cm) adjustable wrench, with its moveable lower jaw, will handle an assortment of jobs. Some come with a locking feature so the jaws won't loosen while in use. Whenever possible, turn the wrench so the heaviest load is against the fixed jaw rather than the moveable jaw.

Cost: $13 to $17

Groove-joint pliers: Groove- or arc-joint pliers are longer and more versatile than conventional slip-joint pliers because they're longer and the jaws open to a 1½ inches (3.8 cm) or greater capacity. And they adjust quickly and easily. The 10-inch (25 cm) size is best for best all-around use.

Cost: $9 to $16

Locking pliers: Vise Grip is a common brand name and the best of the lot. Buy the 10-inch (25 cm) curved-jaw style (Model 10-WR) for all-purpose work, or get a set that includes an additional 5-inch (12 cm) long-nose locking pliers. The release lever is in handle, and on the cheap brands, this is the part that bites. These pliers are particularly handy for gripping rounded nuts to loosen, and they can be used to cut wire or nails.

Cost: $17 to $24 for the set

Long-nose pliers: These are solid-joint precision pliers with vinyl-coated handles. Choose 6 or 8-inch (15 to 20 cm) pliers to reach into tight places or to bend, form, and cut wire.

Cost: $12 to $16

Plumber's helper: This inexpensive invention is nicknamed for the sound it makes when it's working at its best, and it's the only tool that will get you out of certain messes. To unclog a toilet, place the suction cup against the rear exit hole, push down fully, and jerk up to loosen blockage. You may need to repeat several times, or reverse the procedure by pushing air trapped in the cup down the opening.

Cost: $4 to $9

Utility knife: These are available almost anywhere for under $5. For safety's sake, choose one with a retractable blade, and buy a card of extra blades. If possible, try the knife for feel in your hand and test the action of the blade mechanism before you buy. Nothing's more frustrating or hazardous than a jammed utility knife.

COST: $3 to $6

Make the World's Simplest Sawhorse

Cost savings

About $12 a pair if you buy materials, $25 if you use recycled.

Benefits

Recycling scrap lumber and creating a really sturdy work horse.

If you've ever wondered why a carpenter's sawhorse looks so sturdy and stands so well, it's because it was probably made on the job. It's hard to beat something built for utility, and if you build following the directions given here, even the price is right: a pair of these working ponies will set you back only about $12, or nothing at all if you use scrap wood.

MATERIALS

- Circular saw or handsaw
- Measuring tape
- One pound of 8-penny common nails
- Three 2 x 4s (12 feet [3.7 m] long)
- Two scraps of 1/2-inch (1.3 cm) or thicker plywood (at least 18 x 19 inches [45.7 x 48.3 cm])

STEPS

STEP 1: Cut wood pieces. Cut plywood into flat-topped triangles measuring 3 1/2 x 13 1/2 x 19 inches (8.9 x 34.3 x 48.3 cm) tall. Then, for each horse, cut four 32-inch (81.3 cm) legs (from one 12-footer), a 34-inch (86.4 cm) back beam, and two 18-inch (45.7 cm) struts (from half of one of the remaining 12 feet [3.7 m] pieces).

STEP 2: Trim corners. Turn the legs on edge, and trim one corner of each to form a flat face at an angle of 15 degrees. These will join the sides of the back beam at each end. Finish by cutting the ends of the 18-inch (45.7 cm) struts to 45 degrees.

STEP 3: Assemble the saw-horses.
Assemble one at a time. First, nail the legs evenly to one side of the back beam. (They should form square corners when viewed from the side.) Then fasten the remaining two legs to the other side of the beam, taking care to see that the edges of the cut plywood panels determine the splay of the four legs, short of causing the horse to rock.

Set the tops of the plywood triangles against the bottom edge of the beam, and nail them to the legs. Anchor the leg sets to that beam by setting in the two corner struts.

Make Your Own Nontoxic Paint

Cost savings

Approximately $14.30 per gallon of paint

Benefits

Non-VOC (Volatile Organic Compound) paint that you can blend to a variety of pastel colors

Traditional paints and coatings have always been made of two basic materials: milk and hydrated lime. In the days before interstate highways and nationwide rail systems, it was unthinkable to carry something as heavy as paint when it could be easily made locally. Shipping was for manufactured goods such as nails, firearms, and machinery.

One of the best resources for these old-time formulas is the USDA *Farmers' Bulletins*, which were agriculture and home economics publications periodically issued since 1889.

Note: Both of these paints are prone to water-marking even after they've dried, so you may consider covering them with a clear sealer such as Danish oil or another natural or oil-based sealer.

MILK PAINT

This concoction from the nineteenth century uses 1 gallon (4.55 L) of warm milk (skim the cream or use skim milk). Add 1/4 pound (115 g) of slaked lime ("slaked" lime or calcium hydroxide is also called hydrated lime or builder's lime; it is not agricultural lime, which is made from ground limestone). Next, add 2 pounds (900 g) of chalk powder. This is available in the U.S. in 1- and 5-pound (4.55 g to 2.25 kg) jugs as refills for chalk markers. The milk and lime react with each other to form a resin that becomes a powerful binder, helping the paint to adhere to its surface.

Mix enough milk with the lime to bring to a creamy consistency while stirring. Then add the rest of the milk and the chalk powder while continuing to mix. A paint mixer bit chucked into a cordless drill makes the blending easier, but do not stir too rapidly. After a few minutes of blending, apply the paint with a coarse, natural bristle brush. Do not apply a second coat until the first coat has dried completely. Frequent mixing will keep solids suspended in the mix.

Add color with colored chalk dust or commercial pigment powder—or experiment with colored clays or iron oxide (rust powder). The paint will keep several days if refrigerated.

LINSEED PAINT

This formula is more involved. It calls for 1 gallon (4.55 L) of skimmed milk; 1 pound (455 g) of hydrated lime; 1 pint (473 ml) of boiled linseed oil (it must be the boiled kind or the paint will not dry properly); 5 pounds (2.25 kg) of whiting; 4 tablespoons (72 g) of salt; and 4 ounces (115 g) of pine pitch (optional). The whiting is mineral-based calcium carbonate and can be found at an old-fashioned hardware store or a potters' supply, but if you are unsuccessful at that, you can substitute with 6 pounds (2.73 kg) of plaster of paris. (Whiting is the active component of agricultural lime.)

Mix the hydrated lime with a small portion of the milk, then dissolve the pine pitch in the linseed oil and slowly add this mixture to the lime-milk paste. Add the balance of the milk and stir in the whiting powder. Strain the lumpy concoction through several layers of cheesecloth for a better consistency. You can color it with the same pigments mentioned above; some even use berries as a natural pigment. This formula will keep for a day or two, and a gallon will cover about 500 square feet (46.45 square m) of surface.

Make Your Own Wood Stain for Next to Nothing

Cost savings

About $19 per gallon, based on the cost of mineral spirits

Benefits

You control the shade and density, and you make only as much as you need.

Wood stain is not difficult to make yourself, and your wallet will thank you for the effort. First, decide approximately how much stain you'll need for your particular job. As a rule of thumb, check the label of your favorite stain in the store. Then, measure out the volume you need in mineral spirits or paint thinner, which you can often get cheaply at discount outlets.

Next, stir in ordinary mineral-based roof coating, a small amount at a time, until you achieve the shade and density you want. (Be sure to mix in a well-ventilated area and wear safety glasses for protection.) Prepare a wood scrap to test your stain. It's a good idea to use the same kind of wood for your test as you are using for the project, and if consistency is an issue, you should keep a record of how much roofing tar per volume of spirits was used with each sample to achieve that particular shade a second time.

Homemade stain requires frequent stirring. Also, after you've wiped the blend onto the wood surface, you'll need to wipe off the excess. Cracks and crevices can be blown out with an air gun or bumped clear and wiped over to reduce any bleeding out later on in the drying process. Always allow the mixture to dry completely before applying a sanding sealer and any top coats, which will seal in the stain and will also impart a nice depth to the finish.

Preserve and Restore Your Woodwork with Natural Oils

Cost savings

The two liquid polishes will save up to 70 percent compared to store-bought. The polymerized linseed oil is costly, but traditional linseed oils will save about 15 percent over commercial waterproof finishes.

Benefits

Homemade, inexpensive, and natural polishes and finishes

Besides being costly, most premixed, store-bought polishes are full of toxins, petroleum distillates, and unnatural scents that can almost sicken a person at their smell, let alone if they were accidentally ingested.

A few of the *BackHome* staff are woodworking hobbyists, and they've offered this selection of natural homemade wood polishes and preservatives that are safer than most commercial products.

Simple Beeswax Polish

- 1 pound (455 g) block of raw beeswax

This one-ingredient polish is an excellent wood preservative and can be applied easily by hand. Raw settled beeswax is a light, pure wax that softens in the palm with body heat. Rub the softened block directly on the wood, and buff with a soft cotton cloth (a piece of old T-shirt is excellent). If the wax is too hard, it can be rubbed on, and the film softened very carefully with a blow dryer on low heat. Buff out with a cotton cloth.

Spirit Furniture Polish

- 1 ounce denatured alcohol (solvent)
- 1 ounce (28 ml) olive oil
- 1 ounce (28 ml) fresh lemon juice
- 1 ounce (28 ml) pine-gum turpentine (not mineral spirits)

This basic blend stores well. There is no need to use costly extra-virgin olive oil with this recipe, as common cooking olive oil will do. Once you've mix the ingredients, shake in the jar or bottle and rub into the wood with a soft cotton cloth. Polish dry with a clean soft cloth.

Carnauba Wax Polish

- 1 tablespoon (15 ml) carnauba wax flakes
- 1 teaspoon (5 ml) fresh lemon juice
- 2 cups (475 ml) denatured alcohol (solvent)

This is similar to the Spirit Furniture Polish on page 190, but it benefits from the use of carnauba wax, the "queen of waxes," which is derived from the *Copernicia prunifera* palm of northeastern Brazil. Melt the wax flakes in a double boiler (about 180°F, or 82°C), and blend with the alcohol and the lemon juice. Use a soft cloth to apply the polish, then wipe clean with a soft cotton material. Leftover liquid polish can be stored in a jar.

Natural Wood Water Repellent

- 1 quart (946 ml) polymerized linseed oil

This is another straight-up ingredient that protects outdoor wood from water, but the formula must be reapplied annually.

Note: Linseed oil is derived from the seeds of the flax plant, and most of the linseed oil available in stores is either boiled or raw. Raw linseed oil makes a good wood finish for water repellency, but can take a long time to cure—weeks in some climates. Boiled linseed oil dries much more quickly but contains metal driers and petroleum distillates to help it do that. You want polymerized linseed oil, which has been heated anerobically to give it the desirable drying characteristics.

Rag or brush on a thin coat of linseed oil. When it's completely dry (it may take a few days), apply another coat. A few coats should suffice. Linseed oil soaks into wood pores and is highly water resistant.

Make an Heirloom Bread Box

Cost savings

Between $25 and $35 in purchasing a crafted heirloom box

Benefits

Keep breads fresher longer

This yellow poplar bread box will keep breads and pastries fresh and reduce waste through spoilage, which can be a surprising money-saver over time.

MATERIALS

- Table saw
- Jigsaw
- Variable speed drill with 1/4-inch (6 mm), 1/8-inch (3 mm), and 3/16-inch (1.5 mm) bits
- Tape measure
- Drawing compass
- Utility stapler
- Hammer
- Screwdriver
- Router with 1/2-inch (1.3 cm) straight bit (for cutting lid slots)

WOOD MATERIALS

- 1 3/4 x 9 1/4 x 16 1/4-inch (1.9 x 23 x 41 cm) bottom
- 2 3/4 x 9 1/4 x 11-inch (1.9 x 23 x 28 cm) side panels
- 15 3/8 x 3/4 x 16 15/16-inch (0.9 x 1.9 x 42 cm) lid slats
- 16 3/8 x 3/4 x 17 3/4-inch (0.9 x 1.9 x 44 cm) fixed slats
- 1 3/4 x 3 1/4 x 16 1/4-inch (1.9 x 8 x 41 cm) backstop
- 1 1/2 x 3/4 x 17-inch (1.3 x 1.9 x 43 cm) stop
- 1 11 1/2 x 15-inch (29 x 38 cm) fabric
- 1 tube contact cement
- 1 1/4 x 3/4 x 15-inch (0.6 x 1.9 x 38 cm) handle
- 9 No. 6 x 1 5/8-inch (4 cm) cabinet screws
- 4 No. 4 x 5/8-inch (1.5 cm) flat head wood screws
- 64 18-gauge x 3/4-inch (1.9 cm) brads
- 30 1/4-inch (6 mm) utility staples
- 1 1/4-inch (6 mm) dowel or plugs

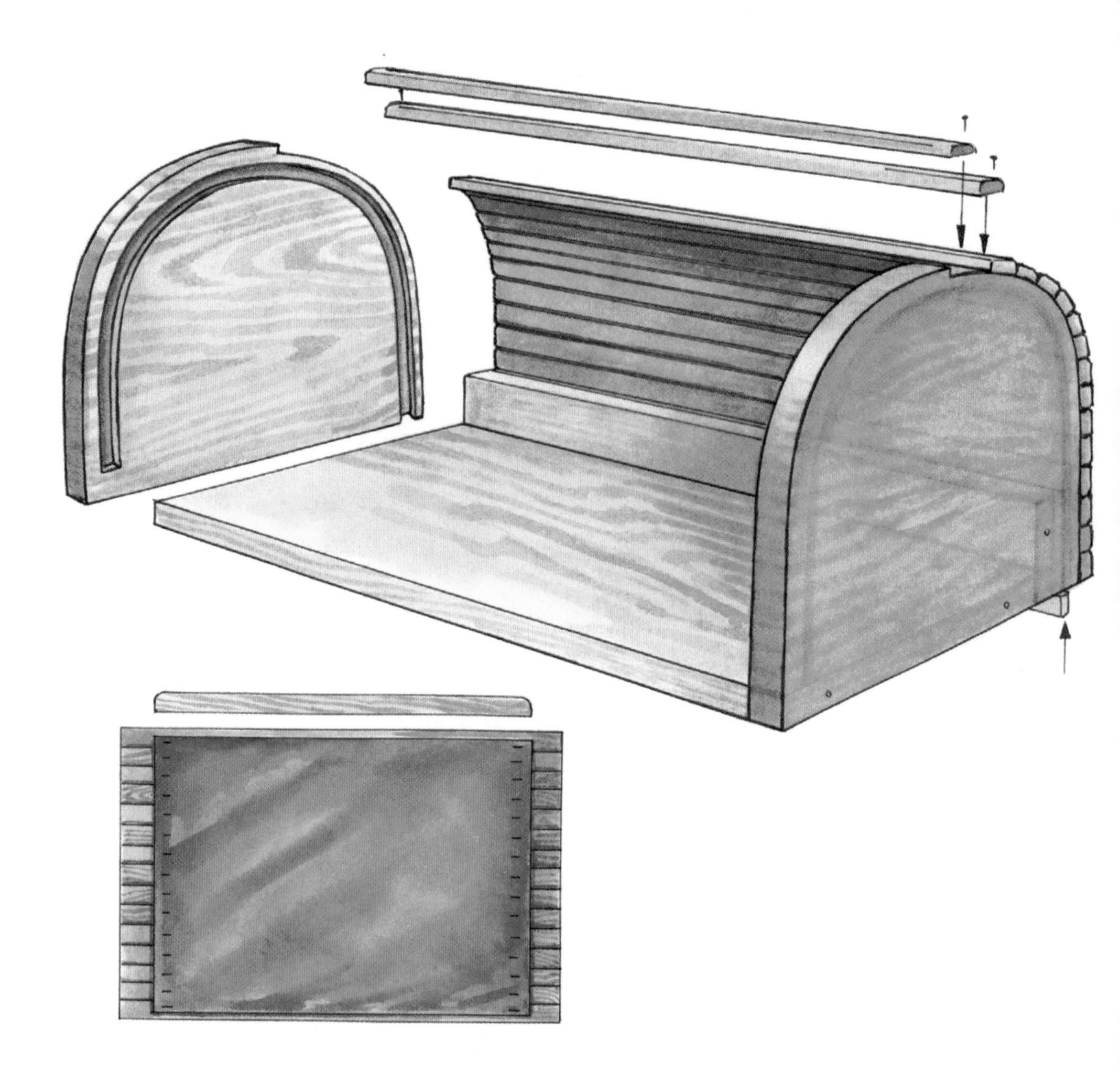

STEPS

STEP 1: Cut wood pieces. Cut the 40-inch (101.6 cm) board to make the bottom piece to 9 1/4 x 16 1/4 inches (23.5 x 41.3 cm). Then cut two pieces to 9 1/4 x 11 inches (23.5 x 27.9 cm) for the sides, taking care that the grain runs parallel to the longest edges.

STEP 2: Lay out curved shoulders. Take one side piece and measure/mark 5 inches (12.7 cm) from the 11-inch (27.9 cm) edge, 4 1/2 inches (11.4 cm) from the edge, and 5 inches (12.7 cm) from the side edge. Mark the two points where these lines intersect, then use them as center points to draw two radii across the face of the panel, using a pencil compass. Remove corners using a jigsaw. Be careful to make a clean cut at the notch where the two different-sized radii meet at midpoint. Use this piece as a template to mark the second side panel for cutting.

STEP 3: Make the bread box face. Choose the two best faces for the outside. Lay them face up on a workbench. Starting at a point $1/2$ inch (1.3 cm) from the outer edge, use a pencil compass to scribe a line to the notch along the inside front (the wider) portion of each panel. Then cut a scrap of wood to the thickness of the notch and use it as a guiding spacer to continue scribing the line along the back.

Next, set your router to a depth of $3/8$ inch (9 mm) and carefully follow the line scribed on each panel to cut a dado groove into the inside faces. The shoulder at the rear of each panel should be about $3/16$ inch (5 mm) wide, and those at the front $1/2$ inch (1.3 cm) in width. Stop the rout at about $3/4$ inches (1.9 cm) from the front edge on both pieces.

STEP 4: Cut the top and slats. These should measure $3/4$ by $1/8$ inch (1.9 x 0.3 cm) thick, or equal to the depth of the notch cut earlier. Make them by setting the fence on your table saw to the necessary width, then running your $3/4$-inch (1.9 cm) boards past the blade, once you've cut them to length. The 15 lid slats should be 16 $15/16$- inches (43 cm) long and the 16 rear ones 17 $3/4$ inches (45 cm) long. After cutting, round the upper edges of each slightly with a piece of coarse sandpaper.

STEP 5: Make the tambour lid. Lay the 15 slats edge to edge, and align them at the ends. Then attach the $1/4$ x $3/4$ x 15-inch (0.6 x 1.9 x 38 cm) handle to the inside edge of an end slat, with four No. 4 x $5/8$-inch (1.5 cm) flathead wood screws set flush from the inside. Cut a piece of nylon webbing or fabric to 11 $1/2$ x 15 inches (29 x 38 cm). Coat the backs of the slats and the fabric with contact cement. Set the fabric in place by centering it side to side and leaving it shy of the front and rear edges by $1/4$ inch (6 mm) or so (trim any excess with a utility knife if needed). Secure the fabric by driving a short staple into each slat along the edges.

STEP 6: Assemble the box. Cut the backstop to 3 $1/4$ x 16 $1/4$ inches (8.3 cm x 41 cm). Butt it to the rear edge of the bottom before fastening it with three No. 6 x 1$5/8$-inch (4 cm) cabinet screws. (The bottom surfaces should be flush.) Fit the tambour lid into the side panel slots and carefully lower the case onto the bottom piece so the front edges are even. Drill two $1/8$-inch (3 mm) holes into the lower edge and one near the top of the backstop to take No. 6 x 1$5/8$-inch (4 cm) cabinet screws. After drilling holes, remove the sides and counterbore the openings to accept wooden plugs. Reassemble the case and sink the screws snug. Cover the holes with thin sections of $1/4$-inch (6 mm) dowel.

To keep the lid from sliding too deeply into the case, cut a $1/2$ x $3/4$ x 17-inch (1.3 x 1.9 x 43 cm) stop and glue it—from the underside—into the tambour slot. A small brad tacked in from each end will help hold it in place. The stationary slats can now be butted together, glued, and tacked in place along the rear edges.

STEP 7: Finish the box. Lightly sand the box and apply a few coats of Danish oil or polyurethane. To enhance the wood with a darker shade, stain it to suit before applying the finish.

CHAPTER 11

DECKS, PATIOS, PORCHES, AND OUTDOOR ROOMS

Decks, patios, and porches are economical extensions of the living space to the outdoors. In this chapter, we'll give you some tips on keeping those outdoor living spaces comfortable and welcoming, using techniques that will keep costs under control.

Clean and Maintain Your Deck

Cost savings

Between $100 and $150 per 150 square feet (14 square m) of deck surface

Benefits

Preparing the wood for finish more effectively than spraying and avoiding the toxic runoff, which damages plants and water supplies

Spruce up your deck without using chemicals. Here, contributor John Wilder explains the process.

Until recently, the predominant method of deck restoration has been to use a store-bought chemical mix and a pressure washer to blow the dirt and mold from the wooden decking. As a contractor, I am here to tell you there is a better, faster and more efficient way: sanding.

By using a right-angle grinder or mini grinder, you can clean and dress the wooden surface quickly and effectively. These tools are powerful enough that you can mount a 7-inch (18 cm) sanding disc on them and work all day without the tool overheating.

The process is simple. Remove the wheel guard and mount a 7-inch (18 cm) sanding backer and a 7-inch (18 cm) sanding disc to the tool's arbor. While operating the tool, tilt it slightly so the forward quadrant of the disc contacts wood firmly (without too much pressure). If you move the disc with the grain it will not raise splinters. Once the weathered wood is sanded, the fresh surface is ready for finishing (no drying time necessary). You'll need a 4 1/2-inch (11.5 cm) backer and sanding pads to sand the railing stiles because of the much tighter clearances. When you are finished sanding down the whole deck, use a leaf blower to get rid of the sanding dust.

While working, be sure to wear leather gloves and protect your face with a dust mask. Safety glasses are a must, and it's a good idea to wear long sleeves and pants.

Finish the wood surface with a semitransparent stain by using a pump-up garden sprayer or a foam pad. In my experience, Penofin, a rosewood-oil based product, is superior to the linseed oil base used in most oil stains. Allow the stain to dry for three days before recoating.

Garden in Winter

Cost savings

$10 to $20 per month saved in buying food at the grocery store

Benefits

Extending the growing season to enjoy fresh food longer in colder climates

A new affordable, earth-friendly trend in seed starting called "winter sowing" gives die-hard gardeners a chance to indulge their plant lust all winter and get a jump on spring. Contributor and year-round gardener Karen Lawrence shares the technique here.

You can create mini-greenhouses from recycled kitchen containers, sow carefully selected seeds, water well, and park the containers outdoors—no matter the weather (really!). Winter is just what seeds need to germinate.

Winter sowing, the brainchild of Trudi Davidoff of Long Island, New York, combines nature's process of stratification (period of moist cold), cycles of freezing and thawing (to loosen seed coats), and natural hardening off (gradual introduction to changing temperatures) with a way to protect seeds and seedlings from heavy rain, snow melt, birds and other animals, soggy soil, and weed competition.

Davidoff hit on her technique while contemplating the problems she faced with too little space for indoor sowing, too little cash for a greenhouse, and an abundance of seeds acquired through Internet seed exchanges. How could she capitalize on the nearly free seed bounty without incurring expense or rolling the dice of indoor sowing in too small a space?

Davidoff tinkered with recycled containers and tested out different seed varieties to see what would work. Here are the basics.

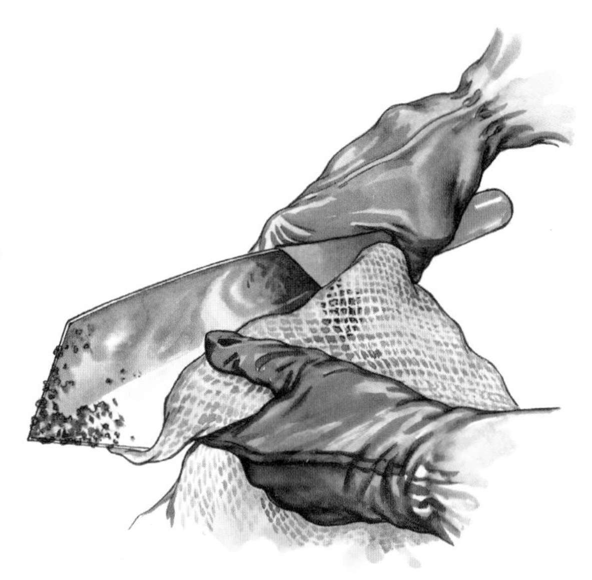

COLLECT CONTAINERS AS DEEP AS YOUR THUMB. Good choices include plastic or aluminum deli or bakery trays with domed clear or translucent lids, milk jugs, 2-liter bottles, takeout containers, whipped topping or margarine tubs (lid equipped with a plastic wrap "window"), or cartons, cups, or meat trays slipped into plastic bags.

SELECT SEEDS. Choose varieties that need or tolerate natural germination conditions and are hardy in your zone. Search seed catalogs for key words and phrases, such as *self-seeds, needs prechilling or stratification, sow outdoors in early autumn or spring while still cool, hardy, direct sow early,* or *tolerates frost.* Other indicators may be plant names such as wildflower, weed, alpine, Russian, and Orientale. Consider varieties that volunteer each year in your area or germinate in the compost pile. Investigate the many seed exchanges sponsored by garden clubs and on the Internet.

PREPARE THE CONTAINERS. The key to winter sowing is creating the perfect microclimate inside each mini-greenhouse. To do that, you need to optimize drainage and air circulation. Cut slits in the top and bottom of each container. A heated knife facilitates clean cuts in plastic. Jugs and bottles can be cut three-quarters of the way around and taped back together after planting. Leave the screw tops off for ventilation. Use scissors to create slits in plastic wrap or baggie tops.

PLANT AND LABEL. Fill the containers with good quality potting soil, sow seeds at recommended depth, water gently, and cover with ventilated lids. Mark each, preferably with permanent marker on waterproof tape. Placing tape on the bottom will keep it from fading in the elements, but don't block the drainage holes.

SET AND FORGET (SORT OF). Place containers outdoors in a convenient, protected spot. The ground, a table, or anyplace where you can observe over the course of the season is

fine. Depending on where you live, locate the containers to receive moderate amounts of whatever winter brings without being blown away or disturbed by animals. Let it rain, snow, and thaw. As long as the flats are properly drained and ventilated they will survive. Check them periodically, preferably on an above-freezing day. Condensation from the sun's heat keeps the soil moist, but if it dries out, water lightly and cover a few of the slits to adjust. If the soil becomes soggy, make a few more drainage holes.

KEEP THE FAITH. New seed flats can be added to your collection throughout the winter, right up until spring begins peeking around the corner. At that point, there will be a burst of growth throughout your little greenhouses. As the seedlings grow, enlarge the slits in the lids. And while the development of seedlings usually coincides pretty closely with local patterns of possible freezes, don't fret if you lose a few of the earliest germinators. Stronger ones almost always pop up in their place.

TRANSPLANT. Treat your mature seedlings like any others—scoop them carefully from their flats, plant them in an appropriately chosen place, water well, and apply diluted fertilizer once the time is right. The biggest advantage of winter sowing is evident once these first seedlings go in the ground. The plants are already hardened off with strong root systems at an early stage, thus are less susceptible to the stresses of the transplant process.

Try Safe, Sound, and Easy-on-the-Purse Herbal Insect Repellents

Cost savings
Several dollars per treatment

Benefits
Safe use of herbal ingredients

Herbs are a safe, convenient alternative to chemical insect repellants and cost a fraction of the price. Here, Cynthia Hummel shares the basics.

Sage: Sage is just one of many common herbs that can repel insects. While sage works best when burned, other herbs and herb-based oils repel bugs when rubbed on the skin or made into sprays.

Citronella: The essential oil of this tall grass has long been used in candles to drive away bugs. You can make a lotion to smooth on skin by combining one part citronella oil with two parts ethyl alcohol.

Cloves: Mosquitoes and moths avoid the scent of cloves, which are the dried buds of the clove tree. Stud thin-skinned oranges with whole cloves. Hang these by string or ribbon from bushes and tree branches.

Eucalyptus and lemon: From the *Doctor's Book of Herbal Home Remedies* comes a convenient insect spray for exposed skin. For every 4 ounces (120 ml) of water in a plastic spray bottle, mix in 10 drops of lemon-scented eucalyptus oil. Shake well before using. Or, combine one part of the oil to four parts white vinegar. **Note:** It is best not to use eucalyptus on children.

Garlic: The same *Doctor's Book* suggests eating garlic cloves for a few weeks before heading into the woods. Campers and weekend barbecue goers can also rub garlic on their skin.

Feverfew: When blackflies are bothersome, soak freshly picked feverfew leaves in some rubbing alcohol and rub them on the skin as a repellent. **Note:** Feverfew should not be taken internally by children under 2, but topically it is okay for children, except for those with common plant allergies.

Lavender: Combine one part oil of lavender and two parts ethyl alcohol as a repellent lotion.

Lemon balm: Fresh lemon balm can be rubbed on skin, rubbed on or crushed over picnic tables, and thrown onto the campfire to keep bugs away.

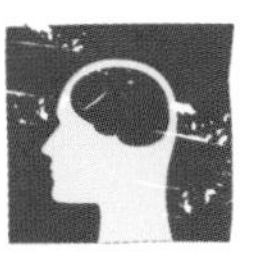

WHEN IT'S TOO LATE

Already bitten? The following herbs can soothe mosquito and other insect bites.

- **Apply a crushed garlic clove to insect bites to soothe them effectively.**
- **Dab the essential oil of lavender directly onto bug bites to lessen itching.**
- **Open up a small portion of a aloe leaf and rub the gel onto the bite. You can prepare some and carry it with you in a plastic sandwich bag.**
- **Another natural bite remedy is made with witch hazel. To a tablespoon (14 g) of baking soda, add enough witch hazel to make a paste, cover the bite for 10 minutes, then rinse it off.**

CHAPTER 12

LANDSCAPING

Maintaining beautiful outdoor space can be an expensive proposition, especially if you have to hire a professional to help. With the landscaping techniques introduced in this chapter, you'll learn to manage your own landscaping chores in a low-cost, all-natural way.

Landscape for Less with Living Mulches

Cost savings

Difficult to quantify, but far less costly than chemicals and labor

Benefits

Pest control, soil stability, and natural energy conservation

Is weeding really a good thing? *BackHome* contributor Gerald Y. Kinro shares the benefits of living mulches.

Overly tidy fields and growing beds? "Not the best situation," says Dr. Joseph DeFrank, weed scientist at the University of Hawaii. "These practices have led to a loss of valuable soil." He references the Dust Bowl of the 1930s that permanently destroyed great portions of North America. Then there is runoff from rain and irrigation. Chemical nutrients can leach out of the plants' growth zone and into aquifers, contaminating ground water. A *living mulch,* he says, can solve most of these problems. A living mulch is nothing more than a ground cover growing with the main crop. Here are some pointers for growing your own.

Plant a cover that grows in your area. Dr. Nathan Hartwig of Pennsylvania State University has researched living mulches for more than fifteen years. He believes that a living mulch must be native to a particular location. In his climate, for example, crown vetch has been used successfully as a cover crop with corn and grains.

Do not plant what you can't control. DeFrank emphasizes that the mulch may compete with your main crop. It may turn into weeds and cover small plants. It may hide rodent damage. He suggests covers that are minimally invasive or are limited in size. Vines are very invasive and may strangle your main crop. Creeping grasses may invade the crops' rows, whereas clumping grasses tend to be less aggressive. As a rule, avoid those plants with rooting at the nodes. Plant something that does not harbor the same pest as the main crop.

Maximize growth during establishment. Take care of your mulch as if it were a cash crop, says DeFrank. The faster it grows, the faster you reap the benefits. Plus you will not want to leave your ground bare for long. Be generous with your seeding. Fertilize during establishment and keep weeds out of the mulch.

CONTROL YOUR MULCH. You may favor nonlethal doses of an herbicide to stunt the mulch. Since this can cause other problems for one not familiar with chemical controls, an option is to mow the ground cover. Some use a "traditional" mulch adjacent to the crop to make management easier.

Living mulches are useful tools for use in erosion control, pollution prevention, pest control, and soil improvement. They are appropriate for small-scale growers and home gardeners. And they look nice. For more information, seek assistance from your local cooperative extension service and the Natural Resource Conservation Service.

Make Mulch from Scratch

Cost savings

About $3.50 per bag of store-bought mulch

Benefits

Getting all the benefits of mulch while recycling natural materials

Contributor Gretchen F. Coyle relies on a homemade "beach mulch" to feed her garden nutrients year-round. But since not everyone lives by the sea, she provides some practical mulch solutions using grass clippings and other common material.

Grass clippings are accessible, high in nitrogen, and break down quickly. In some neighborhoods, the clippings are bagged and set out for trash pickup. No one has ever objected to my helping myself to trash that would be hauled off anyway, and a bonus is that the grass starts to decompose after a day or two in plastic bags. I'm selective, though; I take clippings only from the worst-looking lawns. The grass from finer lawns is most likely saturated with chemicals.

I sprinkle the grass between my rows of vegetables. If spread lightly, the clippings break down in a few weeks. If they are laid down 4 to 5 inches (10.2 to 12.7 cm) deep, it will take longer or they may need to be tilled under a bit.

Other good mulches are newspapers, leaves, and chips from a shredder. Many municipalities have compost, mulch, or wood-chip piles for the taking. Call your town hall or gardening extension service for the location. Then off you can go with a trash can or two in the back of your car.

Do you live near a lumberyard or carpentry shop? Try sawdust and wood shavings, but never use sawdust or shavings from pressure-treated wood. Because fresh wood chips, sawdust, and shavings tend to deplete soil nitrogen as they decay, one method is to spread them on paths between rows, leaving them for a year or so. Then they can be transferred to the garden.

You might be surprised to read that a common mulch, peat, is not a particularly good mulch because it has no nutritional value and tends to get lumpy and crusty on top. Pine needles and oak leaves are not good in vegetable and flower gardens because they tend to make the soil too acidic. However, they are great to mulch around acid-loving trees and shrubs, such as azaleas and rhododendrons, pines, and hydrangeas.

Use your imagination. Natural ingredients for mulching are available wherever you live. Your garden will reap the rewards of your efforts—stronger plants, better yields, and less time watering and weeding.

Conserve Water with Xeriscaping

Cost savings

For normally high-usage areas, a savings of 50 to 70 percent is possible

Benefits

Saving a valuable resource while still enjoying a landscaped yard

Only 1 percent of water on Earth is fresh, and the availability of clean fresh water is dwindling. It has become a commodity, purchased and shipped by pipelines and tankers. Water experts now warn that fresh water will have the same importance in the coming decades as petroleum, and will dominate the shaping of our society. Here, *BackHome* staffer Bill Janes explains how Xeriscaping, or landscaping with a minimum of irrigation, can reduce outdoor water use at home.

We may not be able to regulate regional rainfall, but we can take actions in our own backyards to improve the effects of floods and drought while conserving water. In the drier parts of North America, many property owners—often out of necessity instead of choice—have embraced Xeriscaping, a method of natural landscaping requiring minimum use of irrigation. Xeriscaping reduces the need for herbicides, pesticides, and maintenance. It is aesthetically pleasing because it complements natural surroundings and provides resources for wildlife.

Because Xeriscaping is site-specific, gardeners must learn more about the native plants in their neighborhoods to make good choices about selections for planting. The best sources of information are state agricultural agents, local nurserymen, and public libraries.

In many locations, Xeriscaping will not provide for trees or shrubs for shade and privacy, nor for flower and kitchen gardens. That's where smart irrigation comes in. Drip irrigation delivers the water exactly where you want it, in the correct volume and at the right time. Water savings of 50 to 70 percent can be accomplished if you plan carefully. Even if you live in an area with adequate regular precipitation, drip irrigation pays off in healthier plants and higher yields, plus liquid fertilizers can be delivered directly to the plants' root zones, which need never be dry.

But there are other ways to maximize available water for the garden.

- Build soil with organic amendments such as compost to help retain water and enable plants to develop a wider and deeper root system to better withstand periods of dryness.
- Apply a thick layer of mulch to reduce evaporative loss by keeping the soil cooler and shielding it from wind.
- Position a garden away from water-robbing tree roots and to keep competing weeds under control.
- Regularly check soil moisture 6 to 8 inches (9 to 12 cm) below the surface, and when you *do* find it necessary to supply water, let it run long enough to restore moisture to the lower root zone.
- Use soaker hoses to reduce evaporative loss, and water early in the morning.

Replace Your Lawn with Edibles

Cost savings

Difficult to quantify, but there is savings in replacing jellies and salad ingredients with edible flowers

Benefits

Reduce mowing and fertilizing and provide home-grown nutrition

Edible flowers have nutritional and landscaping value. Contributor Darlene Polachic shares the basics.

Believe it or not, the lowly violet has been termed "nature's vitamin pill." Both the leaves and delicate blossoms of the blue violet are edible, and they're packed with vitamins A and C. In fact, according to one researcher, a half-cup (25 g) serving of violet-leaf greens will fortify you with as much vitamin C as you'd get from eating four oranges and will give you more than the recommended daily requirement of vitamin A.

Violets can be eaten a number of ways—in salads, aspics, jams, syrups, confections, and as garnishes. Violet syrup is made by putting freshly picked violet blossoms into a glass jar and covering them with boiling water. Screw on the lid, and let the solution sit for 24 hours. Strain off the blue infusion, and for every cup (235 ml), add the juice of half a lemon and 2 cups (400 g) of sugar. Bring to a boil, pour into sterilized jars, and seal. This can be served on pancakes and over hot, broiled grapefruit or can be mixed with water over ice for a delightful beverage.

Pansies are another flower whose blossoms and leaves are rich in vitamins A and C. The velvety "funny face" makes a spectacular garnish for summery fruit salads and cream soups. Pansies can also be used to make syrup and to flavor honey.

COMING UP ROSES

Perhaps the most popular edible flower is the rose. The most nutritious part is the rose hip, which contains an incredible amount of vitamin C. Some varieties (*Rosa rugusa*, for instance) contain 20 times as much vitamin C as citrus fruit. In Great Britain during World War II, when imported citrus fruit was virtually nonexistent, children were fed a syrup made from rose hips to provide the necessary vitamin C in their diet.

Rose hips must be gathered when fully ripe. If they're orange, it's too early; dark red, it's too late. Try to pick the hips with the stems intact, since a break in the skin allows light and air to destroy some of the vitamin C. Process the rose hips immediately to retain as much vitamin content as possible.

Wash the rose hips, cover them with water in a glass or enamel saucepan, and cover the pan tightly. Bring the water to a boil, then simmer the hips until they are soft enough to mash. The resulting pulp can be sweetened with honey and eaten as is or be made into jam, jelly, or marmalade. The juice can be strained off and made into a rose syrup in the same way as violet syrup. For marmalade, measure 1 cup (225 g) of packed brown sugar for each cup (235 ml) of rose hip puree. Boil this down to a thick consistency, and store in jars.

Rose petals make a delicious jam for muffins and whole-grain toast. In a blender at low speed, mix 1 cup (50 g) of firmly packed petals, 1/4 bottle of liquid pectin, 1 tablespoon (6 g) of lemon juice, and 2 tablespoons (40 g) of honey until the petals are coarsely chopped. Store in the refrigerator.

MORE TREATS FROM FLOWERS

Nasturtiums

Nasturtiums have long been prized for their nutritive value. A nasturtium leaf is as high in vitamin C as a lettuce leaf, but it is the pickled seed-pods that English sailors took on voyages to combat scurvy. Eventually, the seamen substituted lemons and limes, thus earning themselves the nickname "limeys."

As hors d'oeuvres, stuffed nasturtiums are unsurpassed for flavor and eye appeal. Simply gather fully open, unblemished blooms, and wash them carefully. Allow them to dry before stuffing the centers with small balls rolled from a mixture of cream cheese, well-drained crushed pineapple, and chopped pecans or walnuts.

Calendula

Calendula, or pot marigold, is another valuable food flower. Early Saxons used its leaves as a seasoning in place of salt and pepper. The Romans used both leaves and flowers in salads, preserves, and meat seasonings. Today's chefs use the orange and yellow petals to add a piquant flavor to egg dishes, deviled eggs, and egg salad. The leaves can be pickled or candied.

To candy any flower, layer buds or blossoms in a glass or ceramic bowl with sugar between each layer. Gently pour boiling vinegar (white or cider) over the flowers, and add a piece of mace. Store this for a week before using.

English Primrose

In times past, the English primrose was sought for its healing properties. Dried petals and rhizomes were made into decoctions to clear respiratory passage irritations and into infusions to relieve migraine. The flower is also good to eat. Both leaves and petals of the English primrose may be eaten raw in salads or mixed with herbs to stuff poultry. The flowers make tasty teas, juice, wines, jams, jellies, and preserves.

Geranium

The geranium is often used to lend its distinctive flavor to foods. Consider geranium apple jelly: Simmer sliced apples in water until the fruit is mushy. Strain off the juice, and add white sugar to it cup for cup. Boil the mixture until it reaches the jelly stage, then pour it into sterilized jars with a geranium leaf in each.

Geranium petals can be added to pancake batter to make "flower fritters," and also bring color to fruit breads, such as banana and pumpkin. Chopped carnation, camellia, calendula, and chrysanthemum petals can be used the same way.

Carnations

In the eighteenth century, carnations were used to flavor beer, ale, and wines. You might want to try this carnation sauce, hard to beat over pancakes or warm gingerbread. In a blender, whirl about 5 carnation petals with 1/2 cup (115 g) of yogurt. Place in a bowl and add 5 more whole petals, another 1/2 cup (115 g) of yogurt, 1/4 cup (60 g) of applesauce, 1/2 cup (45 g) of ground almonds, and 1/2 teaspoon (1.2 g) of cinnamon. Stir, and it's ready to serve.

Dahlias

Though the dahlia is edible, at one time people thought the tuber's potential as a dietary staple would match that of the potato. But the dahlia as serious food never caught on. Medically speaking, though, the tuber has served several purposes, as it contains fructose, levulose (a fructose used to test for liver disease), and inulin, a starch useful to diabetics in particular. Today the flower's use as a food is restricted mainly to the petals, which are an eye-catching addition, whole or chopped, to summer salads.

DID YOU KNOW?

Not all flowers are edible. Some, such as foxgloves, four-o'clocks, and lilies of the valley, are deadly. So know your flowers before consuming any part of the plant!

Basic Stonescaping

Cost savings

Between several dollars a month in reduced water bills to several hundred dollars in saved planting

Benefits

Water conservation and reduced maintenance

Complement the garden with pebbles, rocks, and boulders while retaining soil to help plants thrive. Marie D. Hageman shares her stonescaping techniques.

By adapting a method used in ancient Mogul gardens to suit my modern plot, where I breed birds, grow vegetables, and care for flowers, I accomplished two goals: I got rid of our thirsty, labor-demanding lawn and improved the aesthetics of my garden.

Stones, pebbles, rocks, and boulders need no fertilizing, watering, mowing, or replanting. They shine every season. They complement plants, fill bare spaces with immediate beauty, and retain or contain any needed soil.

GETTING STARTED WITH STONE

I drew a plot plan that included such essentials as our purple plum tree, shrubs we wanted to keep, and a fenced-off utility yard for trashcans, a compost pile, and firewood. I expanded all the existing planting beds with graceful, curving lines and defined a couple of raised planting beds in the midst of our flat lawn. These received the topsoil (with grass) that was scraped off the lawn to make room for a couple of inches (5 cm) of small stones.

For low maintenance, I shopped around in stone outlets and garden centers among the tremendous varieties of rocks, crushed stones, and pebbles. I took home small samples so I could see how they looked in their intended place in our landscape under all kinds of conditions: sun, rain, night, day. Stones reflect different lights, in differing intensities, at various times of the day. Some, for example, are blue in sunlight, green in the rain, and silver under the midnight moon.

When I finally selected the stones I wanted, I ordered them to be delivered in staggered loads, each arriving just before I needed to use them.

IMPLEMENTING THE PLAN

First, I traced my new plan on the ground with a shovel and inserted borders to define the sections. I divided the areas with black plastic border strips, rocks, and bricks laid flat in some places and inserted at an angle in others.

Next, I laid a blanket of nonporous plastic on the ground in all areas not reserved for planting beds or for mulch around trees and bushes.

Generally, the more porous the blankets, the more water, air, and liquid plant food they allow through to prevent "sour soil," which can affect tree roots and other roots underneath and nearby. But, alas, the more porous the blanket, the more weeds it allows through. Porous garden fabric reduces weeds; nonporous blocks them.

When our to-be-stoned areas were covered with plastic, I punched holes in the lowest spots to allow rainwater to drain into the ground. Then each area was covered, 2 inches (5 cm) deep, with its own pebbles or crushed stone and raked to form a level surface. My choices of stone for the various sections were pink marble, crushed red shale, river rock, and Jersey Gold pebbles. I also laid a slate flagstone walk around the side of the house and placed large dry-wall

rocks as decorative accents to break up the uniformity of some of the gravel expanses.

The two planting beds made of scraped-off topsoil were held in place by foot-high dry walls of ledge rock. Following instruction given by the stone dealer, I laid the layers so that each rock spanned the meeting place of the two rocks beneath, and all the rocks slanted a bit downward toward the soil they were retaining. Once the planting beds were prepared, I covered them with cedar bark mulch.

ADDING PERMANENT BEAUTY

Next, I added plants to complement and soften the effect of the rocks. These included blue rug juniper creeping over the stone beds with long fingers of green in the summer, which turns to purplish gray-green in the winter, with tiny purple berries. This popular evergreen can thrive in our sandy soil on whatever water nature supplies—or fails to supply. This is one of the secrets of wise planting for unpredictable weather: use plants that can thrive on a wide range of precipitation, a range that is too wet or too dry for other plants.

My garden has survived, beautifully, several years of record-breaking weather—more snow, rain, droughts, and sun than this area ever saw before. Yet my stonescape costs me less in work and water bills than any landscaping I've ever had.

Build a Found-Wood Garden Gate

Cost savings

Between $15 and over $100 for a crafted gate

Benefits

A truly unique handmade gate

Found wood can make a rustic and charming gate. Lori Dzierzek, a rural Virginian, shares how to make this gate for only a few dollars.

You can complete this gate in a few hours using tools and supplies you probably already own. The design differs depending on the materials you use—but you're guaranteed a gate that will keep the chickens in and roaming dogs out.

MATERIALS

- Wood (I use poplar) from a straight-growing tree, 2 to 4 inches (5 to 10 cm) in diameter; consider deadfall and windfall, and avoid wood with excessive limbs
- Cordless drill with bits
- Tape measure
- Hammer
- Lineman's pliers
- Sidecutters
- Aviation shears
- Electric miter
- No. 2 Phillips screwdriver
- Gloves, eye and ear protection

STEPS

STEP 1: Gather your wood. The taller the tree, the more sections you can get from it to make the gate's components. Use the lower and stouter sections for the top and bottom rails and side stiles. Use the narrower upper sections for the crosspieces.

STEP 2: Measure and cut. Measure the height and width of the gate opening (important if fence posts are not exactly plumb). Cut the top and bottom rails to the same length, making sure they'll clear the inside of the opening with a bit of wiggle room. Cut the side stiles to length for the height, so they meet the inside edges of the top and bottom rails (a).

STEP 3: Complete the frame. Lay out the wood pieces on a flat work surface. Position the top and bottom rails against the stile pieces. (This arrangement helps to protect the cut ends from direct exposure to the weather.) Use a 1/8-inch (3 mm) drill bit to predrill holes through the rails and into the stile pieces of the frame. This size is good for No. 8 screws, no matter what their length. Larger-diameter No. 10 screws might require a 5/32-inch (4 mm) bit. Fasten the four main frame pieces together with the decking screws, which come in lengths up to 4 inches (10 cm). With large-diameter pieces, use two screws per joint (b).

STEP 4: Attach crosspieces. Use your imagination, but try to combine a bit of stability with your sense of decoration. For example, position diagonals to connect the top of the hinge stile to the lower corner opposite, to help keep the frame from racking. With that established, place unusually shaped branches or tree parts as you wish on the frame, drilling and screwing the pieces to fit, and using smaller No. 6 screws if necessary. If you want to make a traditional vertical-stile gate, the racking support will have to come from a well-stapled wire covering, and you may want to use denser galvanized hardware cloth in this case.

STEP 5: Attach fencing material. Use either 1-inch (2.5 cm) chicken wire, a heavier hardware cloth as mentioned earlier, or a welded wire with at least a 2 x 3 inch (5 x 7.5 cm) grid. Using gloves and aviation shears, cut the wire to fit the area to be covered. Don't leave any free ends unless you staple them securely to the rails. If you're using chicken wire, cut the unfinished sides about 2 inches (5 cm) longer than needed and fold them over before fastening with poultry staples. I drive the staples at 3-inch (7.5 cm) intervals.

STEP 6: Hang the gate. Use a pair of screw-eye and rigid hook assemblies made for farm use to hang the gate (c). You'll also need a larger-diameter drill bit to install these, and a larger pair of locking pliers. Measure the post and the sides of the gate for clearance and height, and carefully predrill the holes before threading the eyes and hooks into place. If you don't require these heavy-duty hinges on a lighter gate, you can make very serviceable hinges by wrapping heavy-gauge copper or galvanized baling wire in a figure-eight pattern, lopped around the top and bottom of the fencepost, and likewise the supporting gate stile. This works especially well for metal T-post fencing. I usually find a flat stone of the right depth to support the bottom of the gate if I go this route, as it eases the drag on opening and takes the weight off the wire hinges.

STEP 7: Make a gate latch. Heavy-gauge wire bent into a circle or oval serves as an effective latch. Slide it over the post and gate stile to keep the gate closed. If you plan to use electric fencing, replace the wire with an old leather belt, or one cut from a section of rubber tire.

CHAPTER 13

GROWING FRUITS, VEGGIES, AND HERBS

Gardening is a great way to enjoy the outdoors while supplementing your kitchen and pantry with low-cost, freshly grown herbs and produce. Use this chapter as a grower's guide to making the most of your garden space, whether it's simply the edge of your deck or a riot of raised beds.

Save Money and See Results (Really!) Using Vinegar in the Garden

Cost savings

Between $3.42 and $22.40 depending on how you use it

Benefits

Without chemicals and with a single product, kill bugs, fight plant disease, clean tools, and more

Vinegar is a miracle product for the household—and the garden. Here, Pat Kerr shares how vinegar works as a fungicide, poison, and cleaner.

I can't remember when I didn't use vinegar for just about everything, but I started to see big results in my garden when I moved a bottle into my shed. Now I need two containers of vinegar, one bottle full strength (5 percent) and a second bottle diluted 50:50 with water.

Here are some ways I use vinegar in my garden:

Clean garden utensils. Wipe tools clean with vinegar, and if infested with plants, soak them for 20 minutes or longer. Before storing tools, soak them in full-strength vinegar.

Remove lime and hard-water deposits. Wipe pots with vinegar to remove those tough spots; and soak persistent stains in full-strength vinegar.

Clean bird feeders. For hummingbird and oriole feeders, I add sand and gravel and swish this around with vinegar to remove any mold inside. Rinse the feeders well with clear water after cleaning.

Repel ants. Spray vinegar around doors and windows to keep insects out.

Prevent fungus. Diluted vinegar controls the spread of powdery mildew on plants such as phlox.

Damp off seedlings. Plant seeds according to instructions and lightly spray the surface soil with diluted vinegar.

Neutralize soil. For acid-loving plants, use 1 cup (235 ml) of vinegar to every gallon (3.8 L) of water once a month. If potted plants are looking anemic or yellowish, they will benefit from a dose to help release the iron in the soil. When using lime as a top-dressing, rinse of your hands with diluted vinegar to protect your skin.

Beat slugs. Spray the base of plants with diluted vinegar (apply after every rain) to discourage slugs.

Kill weeds. Use 15- and 20-percent concentrations of vinegar (available at independent garden supply houses or from wholesalers) as a weed spot treatment. Intensify "over the counter" vinegar by heating it or by soaking weeds in full sun.

Make a $99 Movable Greenhouse

Cost savings

Up to $1,200 compared to a store-bought greenhouse of comparable size

Benefits

Creating a functional greenhouse with found materials

A few years back, longtime friend Richard Flatau shared with us a story about how he build one of the least expensive greenhouses one could imagine, mostly out of salvaged and found materials. The fact that it was so inexpensive and kept his plants going well beyond the traditional North Central Wisconsin growing season makes his story worth sharing.

In our neck of the Wisconsin woods, having a greenhouse—even a small one—can mean the difference between a short summer's worth of fresh veggies and a *long* season of wonderfully good eating. The value of those early-spring greens—lettuce, broccoli, and spinach—and the late-fall round of peas, turnips, and carrots can't be overestimated when you wish you'd made more of the warm weather while it was at hand.

With that in my mind, I researched all the kit greenhouses and do-it-yourself packages that I could find and came to the conclusion that it would cost between $250 and $650 to put up a bare-bones greenhouse large enough to do any good. At about the same time, I got a $100 offer from my old long-distance company to switch my telephone service back with them. It didn't take long to figure that the hundred bucks would make good seed money for my greenhouse effort, but then I thought, *Wouldn't it be cool if I could build the whole thing for less than the AT&T payoff?*

That way, I'd be setting up a little personal challenge, and I'd have a functional greenhouse at my back door sooner rather than later, when I could actually pay for it.

The gauntlet was down, and I was off in earnest looking for anything and everything I could find to assemble something in a reasonable under-100-square-foot (9.3 square m) range. As it turned out, I didn't have to wait

long. The local lumberyard had a special deal on barn shed trusses: $10 each for 8-foot (2.43 m) x 80-inch (203 cm) tall 2 x 4 frames with a pressure-treated bottom plate. Some quick mental math confirmed that $60 worth would give me a building 10 feet (3 m) long, figuring a standard 24 inches (61 cm) between framing centers for a lightweight structure like this.

The rest was a matter of calculating what I'd need to bring the whole thing together. It didn't take much. Another $10 bought four 10-foot (3 m) 2 x 4s for bracing, two of which I ran horizontally along the inside of the truss studs, just below the lower half of the gambrel roof. The remaining two I screwed diagonally to five of the six studs on each sidewall, to keep the building from racking, or leaning. (By starting at opposite corners I was able to cover more area.)

A single 2 x 6 set beneath the truss peaks tied the top together, at a cost of $4 or so. At the bottom, two 10-foot (3 m) 2 x 4s fastened to the ends of the floor plates completed the box. I framed out a door opening with some rough-cut 1 x 6 cedar, and likewise supported the rear wall by cutting an upright post and joining it midlevel with two horizontal crosspieces and a third brace near the peak just for support.

Then, I waited for one of our community "trash and treasure" days and picked up a discarded but perfectly usable screen and storm door, which I cut down slightly to fit the opening. For glazing, I just bought a 10 x 25-foot (3 x 7.6 m) roll of clear 4-mil plastic and stapled it in place with strips of 1/4 x 1-inch (0.6 x 2.5 cm) wood lath ripped down from old 2 x 4s. I found that a plastic tarp cover is useful, too, for those early spring days that tend to overheat the greenhouse in the warm afternoon following a cool morning. I also slit the rear wall and glued a nylon utility zipper to the seams to provide air circulation.

After a season of use, I feel that storm doors placed at both ends would be even better for cross-ventilation, especially now that I know they're free for the taking.

Inside, I floated another four bucks to buy a 10-foot (3 m) 1 x 10 walk board, which I screwed dead-center onto each of the bottom plates. This gave me a total of ten 22 x 43-inch (56 x 109 cm) planting beds—five to each side of the center walk. I suppose you could prepare the beds any way you please, including lining them with cedar boards or heavy-mil plastic to isolate the treated wood from the soil, but I didn't.

I also didn't bother insulating the foundation area of the greenhouse, because I wasn't ready to commit to a permanent location. As it stands, the structure can be moved as long as its east/west orientation is maintained. It's relatively light, and two or three adults can reposition it. Ideally, it can be set on a few inches of gravel base to let water drain away, or be supported on a course of concrete block, which you could insulate from the outside.

There's little room inside for any heat-storing thermal mass, but a portable heater would take the chill off in a cold-weather pinch.

The greenhouse has been an unbelievable addition to my garden experience. It has provided us with greens and such beginning in April, which is no small feat in northern Wisconsin. A 75°F (24°C) temperature inside while it's 40°F (4°C) and sunny outside is intoxicating. Spring comes slowly up here, and even a small solar space at this price is too good to go unshared. Take it from me—start your planning now and, every spring, you'll be glad you did.

Collect and Use Rainwater

Cost savings

Between $8 and 12 per month, depending on how much water is used and harvested

Benefits

Maintaining a surplus of water for gardening or household use

Collecting rainwater for gardens and nonpotable use helped *BackHome* contributor Kessina Lee-Wuollet avoid the expense of drilling a new well. Here's her story.

When our well ran dry, it wasn't a total shock. One night when I turned on the tap, only gurgles and burps came from the faucet. It was October in northwest Oregon, and locals will tell you that it starts raining about the first of November and rains pretty much until the first of July. Buoyed by the prospect of abundant rain, and fed up with navy showers, we set about constructing our collection system in earnest.

OUR RAIN COLLECTION STRATEGY

We planned to situate three 50-gallon (190 L) food-grade plastic barrels off the back of the house, in a clear spot near the gutter downspout. First, we built a pad for the barrels to rest on by digging up the sod, laying down landscape cloth, sand, and gravel, and setting down level brick. Then we made a low platform using 4 x 4 and 2 x 6 lumber.

Next, using a 1-inch (2.5 cm) hole-cutter, we bored a hole 8 inches (20 cm) from the bottom of each of the end barrels, and two matching holes in the middle barrel. Then we connected them with 1-inch (2.5 cm) PVC pipe and slip-to-thread PVC fittings. We followed the same procedure with two more barrels on the corner of the shop, and one at a corner of the carport to increase our rainwater-collecting surface.

The next step was getting the water from the barrels to the house. One advantage was that we already had the 1,000-gallon (3.8 kl) holding tank as part of our existing water system, so we simply attached a 1/6-horsepower, 15-gallon (56.7 liters) per-minute submersible pump to a garden hose long enough to cover the distance between the barrels and the holding tank.

The hose is attached to an in-line filter placed near the base of the tank. From there, we ran PVC pipe up the side of the tank and plumbed it into another drilled fitting there. To keep the plastic pipes from being damaged, or cracking under the weight of a full filter, we used C-brackets to mount the filter to a solid board. We also installed a filter on the kitchen tap in the house to remove impurities and make it suitable for cooking and washing dishes. Rather than add even more filters to make the water potable, we just purchase gallon jugs of drinking water, or fill jugs at the homes of friends.

Once the rains began, we were able to fill our 1,000-gallon (3.8 kl) tank, use as much water as we needed, and maintain about 500 gallons (1.9 kl) in storage. We try not to let the level in the tank fall below 50 percent capacity unless we have full barrels of water still waiting to be pumped to the tank. In case of a dry spell, we want to have some reserves.

LESSONS LEARNED

The system requires maintenance, of course, and there have been some hiccups. We burned out our submersible pump by leaving it running unattended, and its replacement now has a shutoff timer. (There are also pumps on the market with automatic shutoffs for when demand exceeds supply.) Also, the in-line filter has to be changed about every 1,500 gallons (5.7 kl).

We have a composition roof, which contributes to the amount of debris in our barrels and thus to wear and tear on the filter. A metal roof is optimal for rainwater harvesting, tile and slate are good, and composition is workable though not ideal. Asbestos is suitable only for water not to be used for cooking or drinking.

A simple roof washer alleviates much of this debris problem. The use of an ultraviolet sterilizing filter makes rainwater potable, and a black holding tank prevents algae growth.

The formula for figuring the amount of water actually available for collection is to first measure the area of your roof. It's important to note that you measure the footprint of the roof area you're using for catchment, taking the pitch into account. The rule of thumb is that for every inch of rainfall you'll get 600 gallons (2.2 kl) of water per 1,000 square feet (93 square m) of roof area. To take it one step further and calculate how much water you could collect annually, multiply your catchment area (in square feet) by your local average annual rainfall, multiply that by 600, and divide the answer by 1,000.

This assumes that your system is 100 percent efficient, which no system is. You have to allow for some loss due to roof washing, leakage, and evaporation.

Anyone can harvest rainwater, so long as codes allow it. Harvesting rainwater combines some of the most valued qualities of country living: it's practical, economical, and environmentally sound.

Extend the Season with Row Covers

Cost savings

Approximately $30 to $40 per month

Benefits

Enjoy homegrown vegetables well into November

Gardens provide healthy produce and save you money at the grocery store. With proper planning, you can extend your growing season. A cold-weather garden has always been a simple way to provide onions, garlic, carrots, Chinese cabbage, and a few greens well into November. Without relying on anything as costly or complicated as a greenhouse, dedicated moderate-climate gardeners can make late-season preparations that involve nothing more than working up a few row covers to keep soil temperatures elevated enough to assure satisfactory growth.

What you need is a framework to support a protective cover: something you can easily work around and quickly cover plants up again in chilly temperatures. Stretching the season has its limits, naturally, but even confining yourself to cold-hardy crops will provide you with a generous supply of spinach, chard, radishes, and the various greens most people like without your parting with a cent—which, after all, is almost as enjoyable as the food itself.

MATERIALS

- Six lengths of ½-inch (1.3 cm) PVC pipe
- Remnants of steel rod or construction rebar (slender enough to fit in the plastic)
- Hacksaw
- J-rods
- 6 mil clear plastic polyethylene sheeting
- Two 1-ply wood strips

STEPS

EVALUATE YOUR BEDS. Figure out how long your PVC sections will be based on garden measurements. If the largest bed is 2 feet (0.9 m) wide by 12 feet (3.7 m) long, you'll need PVC sections that are about 5 feet (1.5 m) long. Estimate 3 feet (90 cm) between supports.

STEP 1: Cut PVC. Cut five plastic tubes to length, then drive five of the steel rods about a foot or so into the earth along one edge of the bed with a hammer, each angled inward slightly to match the direction of the arches. Repeat the procedure along the opposite side of the bed, making sure that the rods are spaced evenly and situated squarely at the bed's corners.

STEP 2: Fix arches in place. Slip one end of each pipe over a rod on one side. Bow the plastic tubing until it fits over the corresponding rod opposite. The 6 inches (15.2 cm) or so of steel inside the pipe will be sufficient to hold the arches firmly and keep them from leaning to one side or the other if the fit is snug enough.

STEP 3: Cover the frames. Unravel a roll of 6-mil clear plastic polyethylene sheeting. Allow a few extra feet at the extremities to cover the

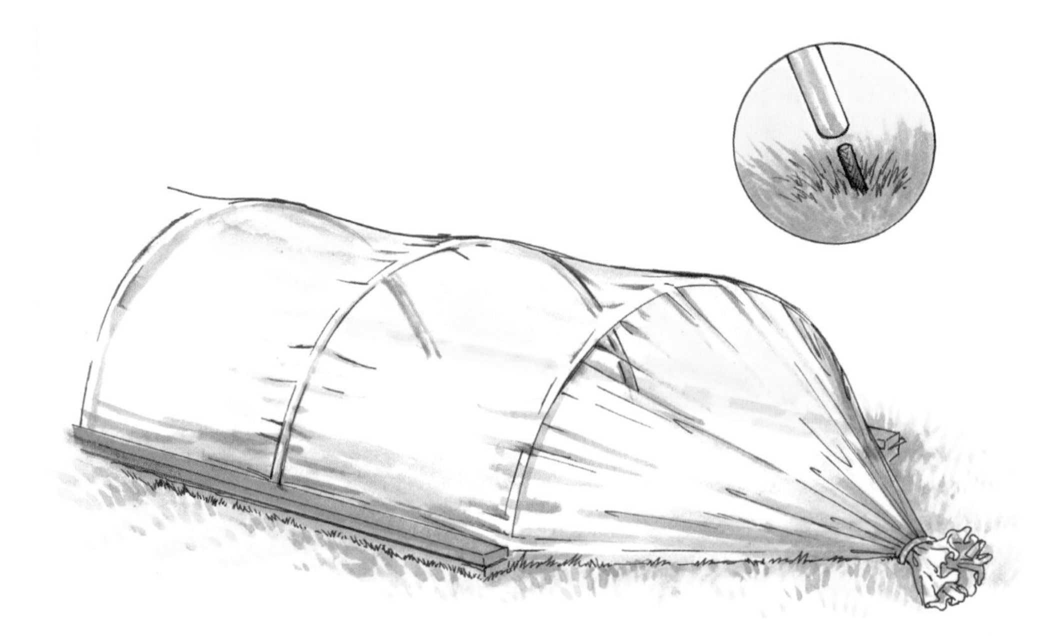

frame ends. Locate a few pieces of 1-ply wooden strip, and cut what's needed to make two 12-foot (3.7 m) lengths. Sandwich one edge of the plastic sheet between the boards, fastening them together with roofing nails to secure the sheeting. Lay that part against one edge of the bed, then spread the plastic over the hoop framework and draw it down to the ground at the other side. Keep that side in place with several more strips of wood secured to the sheeting and held down with a few J-rods. (You can also use steel rod sections bent into a J shape.)

A Word on Cover Material

A lot of people prefer the clear plastic, which you can often got for free from construction sites or businesses that have it left over from packaging (think salvaged mattress covers, boat jackets, and various industrial wrappings). Clear has the advantage of heating the soil and improving cold-weather yields, but it also supports weeds along with the plants.

If you want to get fancy, greenhouse supply companies and maybe some local farm supply outlets might stock a supply of infrared transmitting mulch, which is a wavelength-selective green plastic material favored by professionals and developed to reduce light levels while still delivering the higher temperatures needed to encourage growth.

For gardeners who want to spend money, spun-bonded white polyester fabric, sold under the brand name Remay, among others, is an excellent frost-defying material well suited to extending the season, in both the spring and fall.

Save Money and Time with Sheet Mulching

Cost savings

Between $40 and $160 on purchased mulch, depending on the size of the garden

Benefits

Save time weeding and watering and enhance the soil

Spread free mulch by using techniques shared by contributor Dottie Goudy.

My favorite use of newspapers is for mulching. As a lifelong weed-hater, I had always dreamed of golden globes of cantaloupe peeping from under the umbrellas of their massive vines, spreading a green blanket over a rich brown surface and not a weed in sight! But it was an impossible dream—until I discovered newspaper mulching.

By first putting down a layer, or sheets, of newspapers, you can cover a very large area with only a small amount of grass clippings, straw, hay, or leaves—and the area can be kept *almost* weed-free.

For most plants—melons, beans, tomatoes, potatoes, peppers, cabbage, etc.—the mulching process should begin when the plants are about 4 to 5 inches (10.2 to 12.7 cm) high. Any weeds of a similar height should be removed before putting down the newspaper. Smaller weed seedlings will be killed by the mulch.

Here's how to do it: First, remove the color pages (colored ink has more toxins and does not decompose as quickly as black ink), then spread a layer of newspapers five to six pages thick over the soil. Tear small slits in the paper to slip it around the base of the plants. Then spread a light coating of old hay, or other material, over the papers to hold them in place. Any exposed paper edges, especially in large areas such as a melon patch, may need a little extra hay to prevent the wind from lifting them up and perhaps blowing them away.

Early morning and late evening are the best times for this work. At those hours, the wind is usually calm, and you can put an entire row of papers in place before applying the hay topping. It is very frustrating when a sudden breeze starts flapping and lifting a row of carefully placed newspapers. If a hose is available, a light spray of water on the paper is a great help.

Compost in Your Backyard

Cost savings

From $6 to over $100 compared to purchased soil, depending on the size of the garden

Benefits

Grows healthier plants, recycles yard and garden waste, and reduces use of chemical fertilizers

One of the easiest ways to recycle yard, garden, and kitchen waste is with a compost pile. Composting is simple, and it creates little work for you because the compost pile does all the work, as C. E. Chaffin details here.

The process of composting builds soil in which vegetables can thrive. Ideally, the soil will contain about 50 percent solid matter and 50 percent porous space, to allow for adequate ventilation. Moisture should occupy about half the porous space and air the other half.

Most garden soils are too sandy or clayey. Sand has too much air, clay has too little, and both lack organic material. Adding compost can correct both problems, making the soil a rich source of nutrients for plants. Organic matter contains a number of elements, particularly carbon and nitrogen. Bacteria in the soil use the nitrogen to break down the carbon, feeding the plants. But when the soil is higher in carbon and lower in nitrogen, soil bacteria "borrow" nitrogen for the breaking-down process, and this deprives growing plants, resulting in yellowing leaves.

Composting is a natural process where microorganisms such as bacteria and fungi break down organic material. Gardeners can mimic what happens in a natural setting (like the forest floor) by starting a compost pile. In the pile, you can include almost anything that can decay: grass cuttings, old newspapers, kitchen scraps, vegetable and fruit waste, eggshells, and coffee grounds—but no meat, fat, or bones.

BUILD A BASIC COMPOST PILE

Select an area that's 5 or 6 feet square (2.3 or 3.3 square m). To contain the material, nail together enough boards to make a bottomless container 3 to 4 feet (0.9 to 1.2 m) high. Inside, put a layer of fairly coarse material, such as twigs or cornstalks, to provide ventilation to aid in the breakdown of ingredients.

Build the pile in layers: grass clippings, leaves, weeds, organic garbage, etc. You can add a few pounds of aged manure from a farm. For every several layers of vegetation, add a thin layer of soil. This contains the bacteria that will help break down organic material.

Next, lightly wet the pile until it is moist, not saturated. Moisture content should be 40 to 60 percent, about the consistency of a squeezed-out sponge

Remember that layer of coarse material at the bottom of the bin? This allows air to flow through the pile to speed bacterial action. If you notice the compost is compacting too much, use a stick to punch several vertical holes into the pile all the way to the bottom.

Except for watering, let the heap sit for 2 to 3 weeks. Then turn it, putting the material from the top and sides into the middle. After that, turn it every 3 weeks. When the inside materials look brownish and crumbles on touch, the compost is ready. This usually takes 3 1/2 to 4 months.

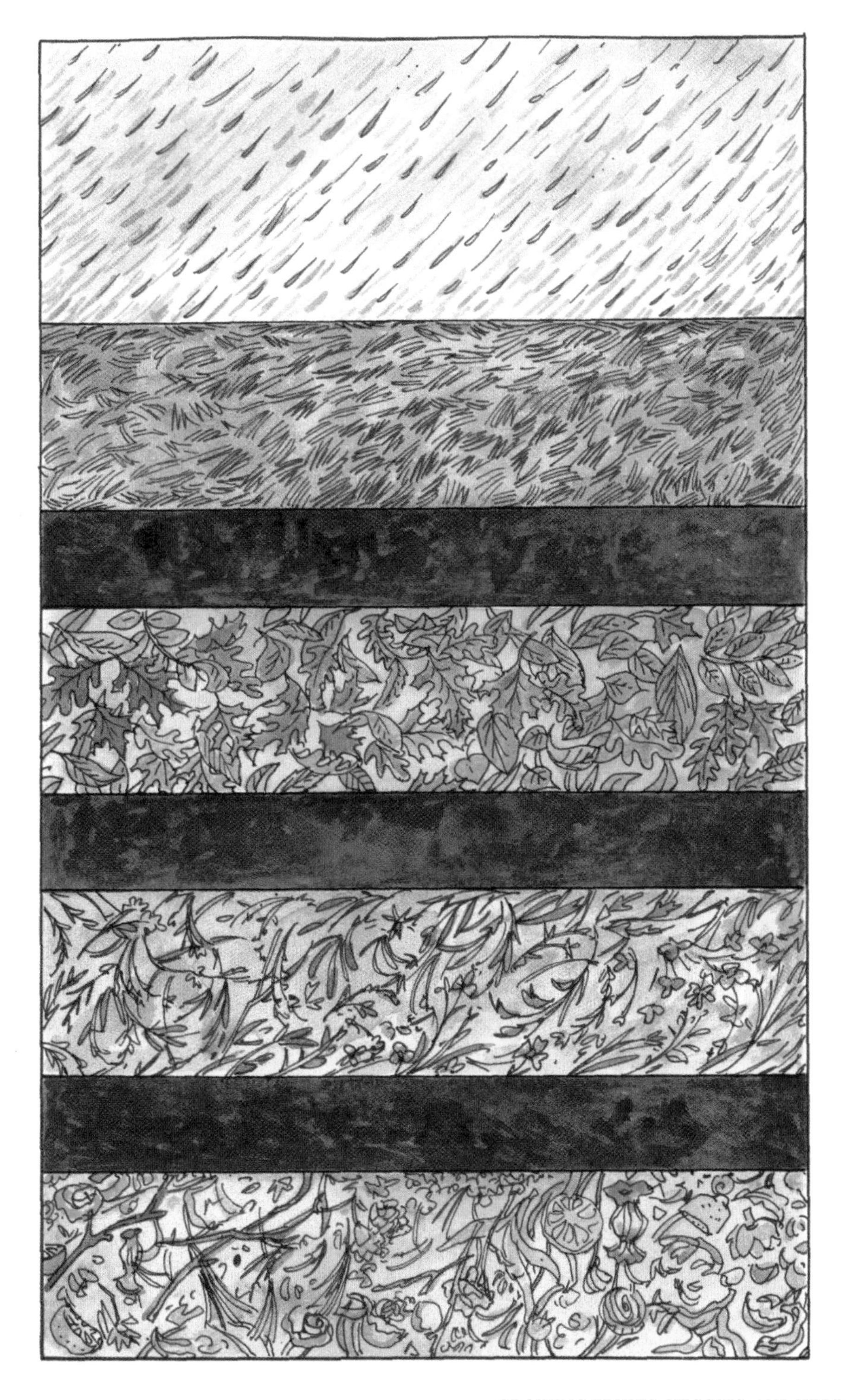

THE 14-DAY COMPOST PILE

If you're starting too late in the spring, or are just impatient, you may want to try an accelerated composting method, developed by the University of California at its Organic Experimental Farm in 1954, that has the pile ready in 2 weeks. The decomposition process is speeded by shredding the materials and mixing them together so the bacteria have many surfaces to work on at once.

First, mix together one part aged (nonprocessed) manure and two parts compost ingredients as listed above.

Using a rotary lawn mower, shred all your scraps, leaves, and grass cuttings. Naturally, you'll want to catch everything in the mower's catcher bag. Put down a small pile of materials and run the mower over it. Repeat the process until all ingredients are shredded into small particles. Better yet, use a power shredder if you have one.

Mix everything together to start the pile. By the second or third day, the middle of the pile should have begun heating, reaching about 160°F (71°C). If it doesn't, add more manure. On the fourth day, turn the heap. Make sure it's warm and moist by putting your hand into the pile. If it doesn't feel moist, add a bit of water.

After a week, turn the pile again, and turn it again after the tenth day. By now the compost should have started to cool.

After the fourteenth day, the compost will have broken down into a dark, rich, fairly crumbly substance. You can use it in your garden right away or let the decomposition process continue.

You can buy drums or cylinders made to rotate easily to turn the compost. Some brands are insulated to assist the internal temperature in reaching the proper levels, and some have baffles to allow for correct mixing when the units are turned.

Tilling or plowing properly prepared compost into the garden soil will correct any carbon/nitrogen shortages *before* you put in your plants. This way, your garden will be ready to perform its task of growing healthy crops.

Build the World's Simplest Compost Bin

Cost savings

Between $55 and $169 compared to store-bought, depending on design and manufacturer

Benefits

Easily assembled from salvaged materials, and can be sized to suit your needs

Commercial drum and bin style compost containers are expensive, and they often don't hold enough working material. A nice slatted box is simple and cost effective if you use salvaged materials. Try plank cedar from old siding, or any weather-resistant wood. Pressure treated 1 x 4s and 2 x 4s will also work, though many people prefer not to invite salt chemicals in the wood from leaching into their garden soil.

MATERIALS

- 32 boards ($3\frac{1}{2}$ inches [8.9 cm] wide)
- Additional scrap wood (for a brace)
- Square level
- No. 8 x $1\frac{5}{8}$ (4 cm) deck screws
- No. 8 x $3\frac{1}{2}$-inch (9 cm) deck screws
- Jigsaw
- Protractor
- Drill and $\frac{7}{8}$-inch (2.2 cm) auger bit
- Sandpaper

STEPS

STEP 1: Gather and cut wood. If you use salvaged pieces, you might have to trim split and nail-pocked ends off a few of the 1 x 4 boards; then cut the good wood into 42-inch (1.1 m) lengths. If using wider boards, cut those pieces into $3\frac{1}{2}$-inch (8.9 cm) widths. You need about 32 of these boards to build the sides of each bin.

STEP 2: Make corner boards. Cut six 2 x 4s to 35-inch (89 cm) lengths. Lay one pair side by side, ends flush. Use a heavy pencil and a small square as a guide. Working from the top, mark lines on the flat faces at these measurements: $4\frac{1}{2}$ inches (11.5 cm), 9 inches (23 cm), $13\frac{1}{2}$ inches (34.5 cm), 18 inches (45.5 cm), $22\frac{1}{2}$ inches (57 cm), 27 inches (68.5 cm), and $31\frac{1}{2}$ inches (80 cm). Create side panels by setting those marked pieces $43\frac{1}{2}$ inches (1.1m) apart, flat on the ground. Place eight of the $3\frac{1}{2}$ x 42-inch (9 x 106.5 cm) boards over them so that the top edge of each one is aligned with a set of pencil marks and all eight are flush against the far edge of the side piece. Leave $1\frac{1}{2}$ inches (3.8 cm) of wood uncovered on the side piece, and a 1-inch (2.5 cm) gap between all the crosspieces. Fasten each board joint with a pair of No. 8 x $1\frac{5}{8}$-inch (4 cm) deck screws. Then repeat the entire process to make a second panel exactly like the first.

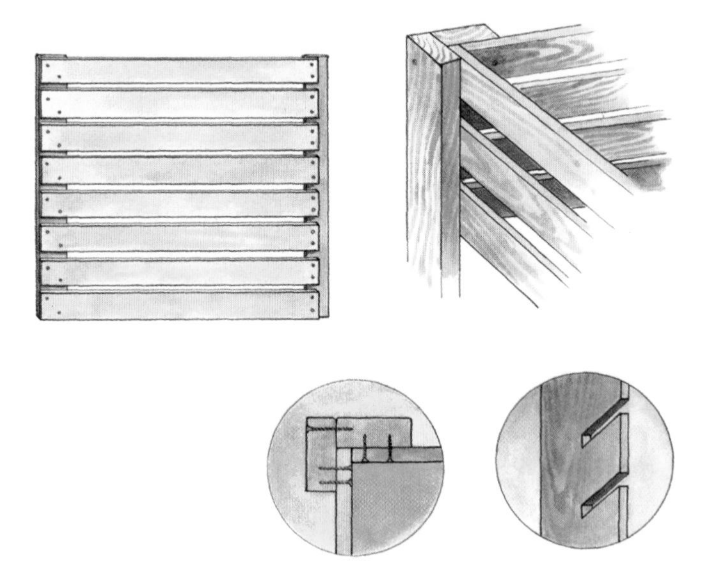

STEP 3: Make the rear wall. The third panel is the rear wall of your wooden compost bin. Follow a process similar to making side panels, with this slight modification. Slat spacing remains the same, but set the ends 3/4 inch (1.9 cm) from the edges of each of the side pieces before fastening the joints. Because you'll be working from the front of the bin, do not use a fixed panel there. Instead, set the last eight slats in slots, which will allow them to be removed individually or in groups, as needed. To do this, cut two 3/4 x 5 1/2 x 35-inch (1.7 x 14 x 90 cm) boards from any wood you have available. Then—beginning at a point 2 inches (5 cm) from one end of one of these—made a series of eight marks along the closest edge, each 3 3/4 inches (9.5 cm) apart. Next, use a protractor to mark eight 35-degree angles at each of those points, measured from the left side of each. To make the slots, extend the angles to 3 1/2 inches (8.9 cm) in length, then draw another line 7/8 inches (2.2 cm) from, and parallel to, each of those. At the ends of each set, use a 7/8-inch (2.2. cm) auger bit to bore a hole completely through the wood.

STEP 4: Make the slots. Cut along the marked lines with a jigsaw to remove the bridges of material. Clean up the cut slots with sandpaper. Then, trace the outlines onto the second board and cut duplicate slots in it. Next, using No. 8 x 3 1/2-inch (9 cm) decking screws, fasten each board to the inside faces of slatted side panels. Slot openings should face outward and be even with the bin's front edges.

STEP 5: Create a temporary brace. To brace the boxes without making them permanent fixtures, drive a few No. 8 x 3 1/2-inch decking screws through the abutting sidepieces at each corner to hold the rear together. Then screw two 3/4 x 2 1/2 x 46 1/2-inch (1.7 x 6.4 x 118 cm) strips of scrap across the top and bottom of the side panels up front. Remove screws to disassemble and move the bin. Slip the front slats into their slots to ensure proper alignment. If the slats don't come in and out easily, trim them down or refasten the two front cross braces after making adjustments.

Learn Seed Saving Skills

Cost savings

Saving money on your favorite varieties by harvesting your own seeds

Benefits

Rescuing heirloom varieties

Long before you could buy packets of seeds, gardeners saved the seed of their best plants, sowed them the next year, and improved the species. Here, contributor Beverly K. Metott walks us through the art and science of seed saving.

Gathering seeds from plants makes good gardening and economic sense. The cost of an individual seed packet is well over $1, and there are not a lot of seeds inside that packet. When you consider multiple varieties, costs can quickly add up. Go to a nursery in the spring to purchase sprouted equivalents, and you're talking about handing over quite a bit of cash for something you could grow yourself for free.

SEED BASICS

Every year you can find me out along the walk and in the garden, deadheading blooms that hold precious seeds for next year's crop. The best time to gather seeds is when the flower heads turn fully brown, before they split to release seeds into the soil. Timing is everything when seed collecting. Also, only nonhybrid plants should be collected for the following year's crops. The seeds of nonhybrids will produce genuine plants that are very much like the parents. Open-pollinated and self-pollinated varieties are dependable and bring the best results for seed savers. A good place to start is with annuals or biennials: plants such as peas, beans, tomatoes, or—if you're considering flowers—poppies, nasturtiums, marigolds, etc., keeping in mind that perennials are usually propagated through divisions or cuttings.

Keep in mind there are three main pollination methods: airborne, insect, and self. If you want the seed to have the same genetic composition of its parents, it must be pollinated with pollen from the same variety. If a plant is airborne pollinated, there can't be other varieties within a mile (1.6 km) shedding pollen at the same time or else some of the harvested seed will result from a cross between the two varieties. The closer the varieties are located, the higher the percentage of crossing.

To counter this, grow one variety of a plant in your garden to decrease the risk of cross-pollinating. For instance, grow only one type of spinach when expecting to save the seeds. Two different types run the risk of cross-pollinating. If you must grow two varieties, consider self-pollinated crops.

Self-pollinated crops are the most ideal plants from which to gather seeds because the pollen is transferred directly and automatically to the stigma within the flower. If two or more varieties of the same plant are in the garden, separate them with a row of an entirely different plant species.

COLLECTING SEEDS

Gathering seeds is easy and, basically, a matter of timing. The trick is to select seeds from your best plants. Know your plants, and examine them throughout the growing season. Check for color; the plant's overall vitality; its flavor, smell, size, yield, and condition. Don't save seeds from a plant that did not fare well. When you go outdoors to begin collecting, consider the whole plant and its successes and failures throughout the growing season. If you have multiples of the same plants and may forget which ones looked the best, mark these plants with stakes so they will be easy to go back to later on.

Once fall approaches, survey the plants you've staked and begin checking for blooms (or heads) that have started to dry up and turn brown. Here are the seeds that are ready to be collected. Some heads may even begin splitting open as the plants prepare to do what nature intended—release their seed into the air. If you gather seeds too early, they may not reproduce and/or the job of separating the seeds from the flower head will be that much more difficult. Wait too long, and they will have fallen to the ground below. Observation is definitely the best method for gathering flower seed.

After collecting dried flower heads, invert one variety at a time over a clean container and shake the flower head against the sides. You will see the seeds fall onto the bottom of the container. Transfer the seeds to a paper towel or newspaper for about two weeks and put them in a safe place away from kids and animals. Write the name of the plant on the paper so no confusion takes place. Once dried, transfer the seeds to an envelope labeled with the flower or vegetable type, and seal. Store the seeds in a cool, dry place until spring. For the most viable seed, it is best if the seeds are used the following year.

Note: Not all varieties are easily gathered by this method. For instance, a California poppy doesn't mature inside a flower head as a marigold does. Talk to other gardeners about their experiences, consult a nursery professional, or refer to a university extension for guidance.

Saving seeds is one small step toward self-reliance, plus you will develop a more thorough understanding of the whole process of a plant's life, not to mention a better eye for quality plants. One thing is certain: saving seeds brings the life of a plant full circle.

Know the Ins and Outs of Container Gardening

Cost savings

By growing your own vegetables, you save about $30 and $70 a month; doing your own composting you will save additional expense; otherwise container growing saves considerable time in weeding

Benefits

Weed- and chemical-free gardening, larger yields, portable plantings

Container gardening runs the gamut from low-tech planting in pots and bottomless forms such as used tires to high-tech systems employing irrigation and supplemental lighting. In between is a world of possibilities, opening the door to personal food production for almost anyone. *BackHome* contributor Mrs. John Bayles shares container gardening pointers here.

My last year of "normal" gardening came hard on the heels of four sneaky nanny goats. The green beans were producing in bushels, the pumpkins were nearly ripe, and the tomatoes were a chorus of red perfection. I had tilled and watered and worked hard over this garden, and I cried when I saw what the goats had done overnight.

The real problem was logistics. My garden was so far from the house that I couldn't see when farm critters circumvented the fence. Since drastic measures seemed necessary, I chose to plant everything in pots and move the garden closer to the house. This requires only a hand shovel, soil, and some form of container, and works wonderfully for anyone who's short on space. You just line up the pots on your deck, balcony, or terrace, fill them, water them, and watch the plants grow!

CONTAINER GARDENING BENEFITS

You can virtually eliminate weed seeds and soil pests in container gardens by sterilizing the soil before potting it—and you'll never have to touch chemicals. Simply place soil 3 to 4 inches (7.6 to 10.2 cm) deep in baking pans and cover them with a layer of aluminum foil. Heat the soil to between 180 and 200°F (82 and 93°C) for approximately 30 minutes. Do not exceed 200°F (93°C) in the baking process.

Any weeds that grow take only a few minutes to remove. Another bonus: watering efficiency. The water you give your plants in pots goes straight to their roots and doesn't soak wastefully into the ground. You can't grow an acre like this, but you can plant the seeds a little thicker than usual since the roots can sink down the full 1- to 2-foot (30 to 60 cm) depth of your pot.

My pots sit on the south side of my house, but if a storm is coming, I can move them under shelter until it passes. And early frosts, too, can be bamboozled by moving the pots to shelter on cold nights.

As for container variety, they come in all shapes and sizes: I use large terra-cotta pots, half whiskey barrels, and old tires. Yes, the tires are incredibly ugly, but they make the best growing pots I've found. They're sturdy, the black color absorbs the sunlight, and—best of all—they're free at tire stores, which actually encourage you to load up. Stack the tires three or four high, fill them with dirt, and plant all your root crops in them. At the end of the season, a good, swift kick or two, and you'll have potatoes and carrots lying all over, ready to be picked up and stored for winter.

COMPOST IS THE KEY

Pot gardening is most successful if you make compost to fill the pots. (Refer to page 222 for instructions on making your own compost.) Commercial potting soil is the second best material, but it is expensive. You *can* just use good black topsoil if you can find it, but be sure to mix vermiculite or peat moss in it for aeration. The easier it is for your plants to dig their toes into the soil, the more nutrients and water they will soak up.

Compost can be made out of table scraps, yard trimmings, and livestock droppings. I have found that regardless of what materials I'm composting, a bale of straw makes it nearly foolproof. Mix the food scraps and other materials with an equal amount of straw, and mix well. Keep it in the sun and watered, and now and then stick your hand in the center to see how hot it's getting. It should be no hotter than 140°F (60°C), which is about the upper limit of what your hand can take. If the compost is hotter, stir it more often.

When the compost looks like black dirt, put some in a jar. Fill the jar with water, screw the lid on tight, and let it sit in the sun. After a day, the water should be clear; if it's green or murky, the compost is too raw and will burn your plants. Add some wood ash, if you have some, to your heap; mix, water, and wait a week to test it again.

BIGGER AND LONGER LASTING

The roots of plants grown in pots get warmer than roots in the garden plot, so you'll need to give the pot gardens more water than you would a conventional garden. But a word of warning: the high nutrient content of the compost, frequent watering, lack of weeds, and deep rooting can result in huge plants. If growing sweet corn, for instance, you'd better keep the tall stalks out of high winds.

Movable pots have given rise to a new pleasure for me: indoor gardening. By January my cantaloupes are blooming in the kids' room, the cucumbers are flourishing on the living room windowsill, and the tropicals—avocados and mangoes—are growing behind the woodstove. No longer do we have to watch our hard work die in the fall. Our tomatoes continue to produce through the winter, and then we set them back out in the summer for more fresh air and sunshine.

Even though I have the land for more extensive gardening, I grow plants in pots because it's less work. I have more control over weeds, insects, and weather, and the results are more satisfying. But container gardening is also perfect for those who have little land or even those who live in city apartments. Colorful flowers on the balcony are lovely, but think how much more you'll enjoy harvesting vegetables.

Create a Raised Bed Garden

Cost savings

Between $20 and $90 per growing season in insect and weed control

Benefits

Raised bed gardens are productive, functional, and convenient

Raised beds are accessible, economical, and easier to care for than field gardens. Here, veteran gardener Nancy Reiss discusses the practicality and costs of raised bed gardening.

After more than a decade of digging in the dirt, I have evolved from an eager neophyte into a lazy, middle-aged gardener. About five years ago, concerned with my limited time and decreasing energy, I became discouraged with the condition of my raised beds. Built of recycled boards that had rotted beyond repair, they were chock full of weeds and had wasps nesting in the wood. I considered rebuilding with wood, but I rejected this idea because of the expense and labor. Also, in avoiding the rotting problem, I didn't want to hazard the toxicity of wood preservatives or pressure-treated landscape ties, which contain chromated copper arsenate.

Meanwhile, my regular gardening space was becoming too large for me: I planted too much in spring, became overwhelmed with weeds in summer, and neglected necessary watering.

Then I read Mel Bartholomew's book *Square Foot Gardening*. I was convinced I should learn to garden more intensively. I decided to rebuild my raised beds out of concrete blocks.

BUILDING THE BEDS

After planning a series of 9 square foot (0.8 square m) beds, I bought ten concrete blocks for the first bed—hollow-core seconds, 8 x 8 x 16 inches (20.3 x 20.3 x 40.6 cm)—that I got for 65 cents each. Before laying the blocks out, I measured the perimeter with stakes and string, roughly leveled the soil with a spade and rake, and put heavy-duty (6 mil) black plastic under the entire bed, poking holes just in the center to allow for some air circulation and drainage.

I assembled the first course: three blocks on two sides, framing two blocks on the other two, making each exterior side 48 inches (120 cm). I filled the bed with soil and realized I hadn't figured the width of the blocks correctly to come up with 9 square feet (0.8 square m) of interior space. So, rather than follow the book plan, I enriched the soil with acidic organic materials, planted the whole bed with strawberries, and mulched it with pine needles.

I built three beds that first fall—the first was a "tester," and I discovered that I needed to revise Bartholomew's dimensions to suit my own space. The other two beds were built according to the diagram on page 233. I made each bed of fourteen blocks, with four-block and three-block sides, making a planting area that totaled 16 square feet (1.5 square m). I divided each plot by threes, marking a grid pattern with string attached to stakes hammered into the ground. My squares were 16 inches (40.6 cm) each side.

Where I couldn't rake the ground level, I used flat stones or wood wedges under the blocks so they butted together tightly. I added lime, wood ashes, and well-aged manure in old sawdust to the soil in my new beds.

Beneath the beds, I advocate total coverage with black plastic, punched with drainage holes. The increased heat allows earlier gardening and a lengthening of the season. The water retention and extra warmth help balance the wicking effect of the concrete blocks. Although they need extra moisture, the warm-weather crops thrive surrounded by black plastic.

THE FIRST YEAR

In my eagerness to try my new beds, I planted peas on March 17. With our cold, wet, New Hampshire spring, they didn't get the early start I'd hoped for. Deciding the bed had become too waterlogged because of the plastic underneath the soil, I built another bed, but cut 2 square feet (0.19 square m) out of the center of the plastic, and planted more cool-weather crops: peas, spinach, and lettuce. Along with these two beds, I still had two old wooden raised beds. Then I built two more block raised beds outside the fence.

The cool-weather, moisture-loving crops fared best in the block bed with total plastic underneath, because it retained water better. The bed in which the plastic was partially cut out dried out much faster.

In dry weather, the concrete blocks seemed to wick water out of the soil, hastening wilting. But after a rain or a good soaking, they retained moisture. Also, I could fill the hollow cores with water, and plants thrived in the gradual moisture provided. Though it wasn't a dry summer, I did have to water the concrete beds more often than I did my original garden. Weeding was a breeze with the block beds. Every part of each plot was within an arm's reach. The soil stayed soft, and the weeds' roots offered little resistance when yanked. With only an hour a week of bending over the beds, I could stay on top of the weeds. For the first time in more than a decade, the garden didn't turn into a jungle.

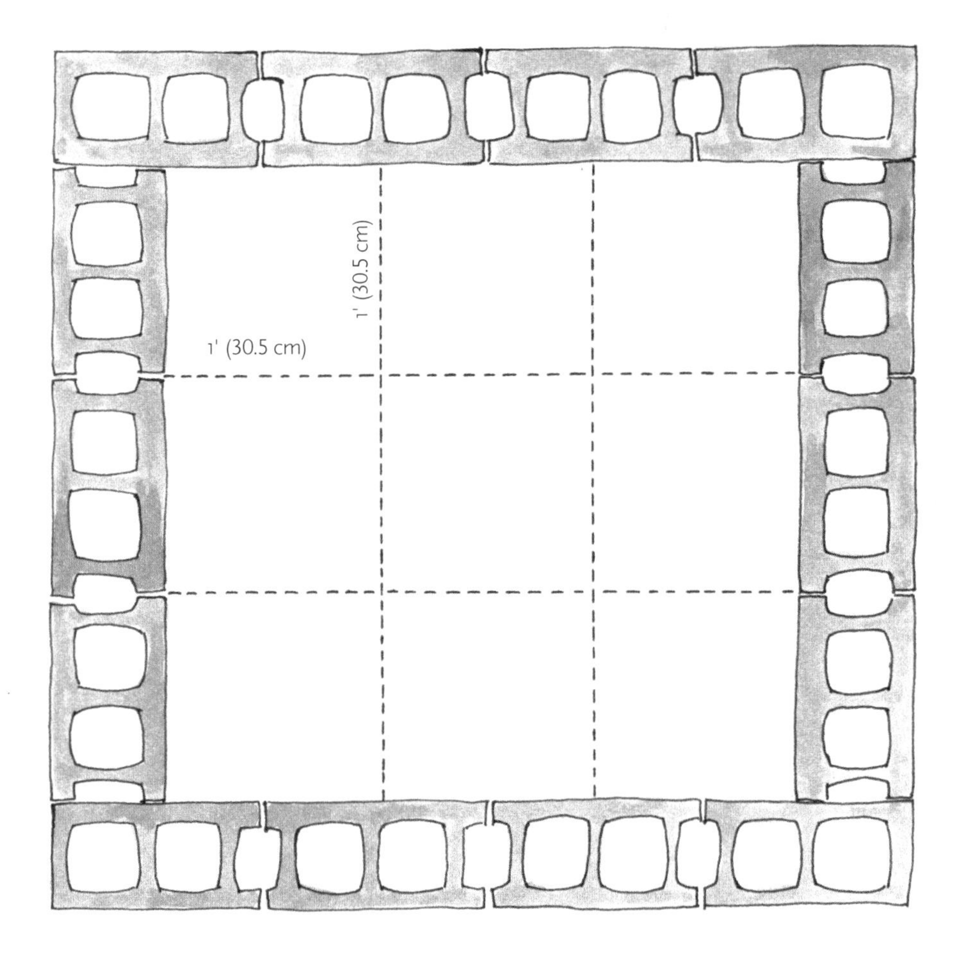

I also learned to store small tools and watering cups in the hollows of the blocks. With supplies at hand, I can weed, feed, and water one bed in a matter of minutes. Another handy use of the hollow cores is to plant them with herbs or marigolds. Invasive herbs such as mint are contained, yet will come back year after year.

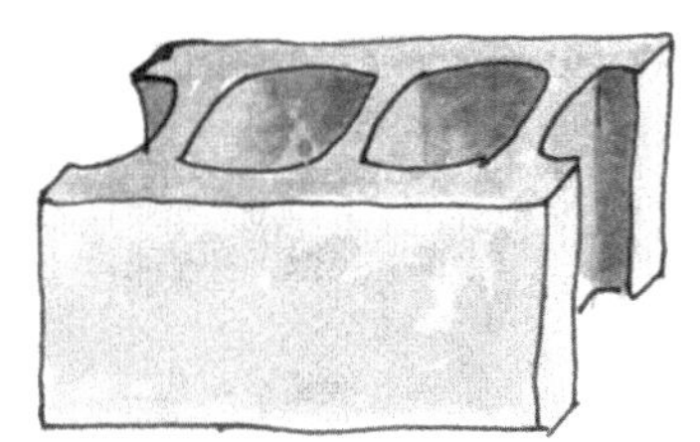

So what does experimentation with my new raised beds indicate? I rate square-foot gardening with concrete blocks a success. The time saved on weeding more than makes up for any extra watering. I would recommend this method for all shallow-rooted moisture-loving vegetables.

Use Beneficial Insects for Garden Pest Control

Cost savings

Between $19 and well over $50 in saved pesticide expense

Benefits

Reduces or eliminates use of insecticides and creates a flowering garden

Although there are plenty of insects we want to keep *out*of our gardens, other bugs are *useful* in the garden. Beneficial insects play a vital role in seeking and destroying pests and they're an alternative to expensive pest control chemicals. By providing a variety of nectar and pollen, you can attract and nourish beneficial insects—for free. Here, contributor Kris Wetherbee explains what to plant to entice allies in your fight against the bad bugs.

Get familiar with flower families that attract the good bugs, and plant plenty of them.

Umbel family: The favorite foods for many beneficial insects can be provided by growing flowering members of the umbel family or the composite/daisy family. Parasitic wasps (tiny wasps that don't sting) love small flowers with easy-to-reach nectar, and ladybugs, lacewings, hover flies, and other predator insects will also feed on their pollen. While dill is a favorite, many other umbel family members are great for attracting the good bugs. Just think of other flowers that are umbrella-shaped. Fennel, parsley, coriander (cilantro), and sweet cicely all will provide food for beneficial insects and culinary herbs for you. Natives such as Queen Anne's lace are also good.

Composite family: Larger predatory insects such as ground beetles, rove beetles, and others visit larger flowers for food, and many of their favorites are members of the composite family. Any daisylike flower with petals circling an easy-to-reach pollen-laden center will provide food. They do double duty as cut flowers, and marigolds are one of the best companion plants for the vegetable garden.

Mint family: Flowering members of the mint family also attract beneficial insects, including the tachinid fly. (They look somewhat like houseflies but are voracious destroyers of Japanese beetles, Mexican bean beetles, grasshoppers, and more.) Spearmint, peppermint, lemon balm, bee balm, and thyme are powerful lures for these insects. As a bonus, you can enjoy sipping mint tea while relaxing in your garden.

Corn and epazote: Epazote is a popular spice used in Mexican cooking. If you forget to harvest your broccoli and it has begun to flower, leave it alone! This, too, will provide a haven for beneficial insects.

Nature is diverse, and our gardens should be the same. Have flowers blooming all season long for beneficial insects to feed on. Sow plants such as dill and cilantro several times. Plant native grasses as winter habitat for ladybugs, assassin bugs, and others. Hydrangeas and goldenrod are a favorite of the soldier beetle. (It looks like a firefly). These insects dine on grasshopper eggs, caterpillars, and cucumber beetles.

Bug-Friendly Behavior

Beneficial need plants to eat, but they also require water, shelter, and protection. Give them a shallow birdbath lined with pebbles. Add decorative rocks to the garden for protection. And do some research to make sure they're not invasive to your area.

Keep in mind, products made to kill insects don't discriminate, killing both unwanted pests and beneficial insects. Those bug zappers zap more than moths and mosquitoes. Many beneficial insects, including the tiger beetle, are also attracted to them.

Finally, if you want to attract beneficial insects, avoid pesticide use (even organic products). For example, rotenone kills beneficial insects and also bees. However, *Bacillus thuringiensis* and neem oil are safe because they affect only pests that eat foliage of a sprayed plant.

Beneficial Flowers to Grow

Flower	Type	Sun	Moisture	Height	Bloom	Insect
African Daisy	Annual	Full	Light	4–12 inches (10–30 cm)	Summer	Lacewings, minute pirate bugs, ladybugs
Alyssum	Annual/ perennial	Partial to full	Light	8–10 inches (11–25 cm)	Spring to early summer	Tachinid and hover flies, chalcids
Angelica	Biennial	Partial to full shade	Moist	4–6 feet (122–183 cm)	Summer (long season)	Ladybugs, wasps, lacewings
Bee Balm (Monarda)	Perennial	Partial to full	Regular to moist	2–4 feet (61–122 cm)	Summer	Minute pirate bugs, lacewings
Calendula	Annual	Full	Regular	1–2 feet (30–61 cm)	Late spring, summer	Minute pirate bugs, lacewings
Candytuft	Annual/ perennial	Full	Regular	6–18 inches (15–46 cm)	Spring to early summer	Hover flies
Chamomile	Annual/ perennial	Full	Light to regular	12–30 inches (30–76 cm)	Late spring, summer	Hover flies, ladybugs, lacewings, parasitic wasps
Coreopsis	Annual/ perennial	Full	Light	6–36 inches (15–91 cm)	Late spring, late summer	Minute pirate bugs, ladybugs
Crocus	Perennial	Parial to full	None	Up to 5 inches (13 cm)	Late winter, spring	Parasitic wasps
Dill	Annual	Full	Light	Up to 4 feet (122 cm)	Summer	Parasitic wasps, lacewings, ladybugs, tachinid and hover flies
Echinacea	Perennial	Full	Average	2–5 feet (61–152 cm)	Late summer, autumn	Ladybugs, minute pirate bugs, beetles

Flower	Type	Sun	Moisture	Height	Bloom	Insect
Elecampane	Perennial	Full	Moist	3–5 feet (91–152 cm)	Summer	Minute pirate bugs, ladybugs, beetles
Feverfew	Perennial	Full	Light to average	2–3 feet (61–91 cm)	Summer	Ground beetles, hover flies, ladybugs, lacewings
Goldenrod	Perennial	Full	Light to average	3–5 feet (91–152 cm)	Summer to midautumn	Big-eyed bugs, minute pirate bugs, ground and soldier beetles, lacewings
Sunflowers	Annual	Full	Light to average	3–10 feet (91–305 cm)	Summer to midautumn	Lacewings, ladybugs, beetles
Sweet Woodruff	Perennial	Partial to full shade	Moderate to moist	6–12 inches (15–30 cm)	Late spring	Lacewings, hover flies, minute pirate bugs
Yarrow	Perennial	Partial to full	Light to average	3–5 feet (91–152 cm)	Midsummer to early autumn	Ladybugs, parasitic wasps, tachinid flies
Zinnia	Annual	Full	Moderate	1–3 feet (30–91 cm)	Summer to late autumn	Flower bugs, ladybugs, lacewings

Companion Plant for Gardening Success

Cost savings

Boost your crop yield and save $25 or more on pesticide expense

Benefits

Repel harmful insects in your garden

Basil and tomatoes make zesty allies in sauces, on pizzas, and in salads that it's hard to imagine cooking tomatoes *without* adding a sprig of basil. But did you know that growing basil and tomatoes together in the garden will improve your tomato crop?

Companion planting is a holistic gardening technique based on diversity in the garden. Rather than planting segregated rows and plots of each vegetable, you intersperse varieties of plants to take advantage of their natural relationships. This simple gardening technique can help you save money by increasing your yield and reducing the need for insecticides. Here, Kris Wetherbee fills us in on the ins and outs of companion planting.

Companion plants can help improve your garden yields in three ways: They can repel insect pests and provide nourishment, which we'll explain here. They can also attract beneficial predatory bugs. (See "Use Beneficial Insects for Garden Pest Control" on page 111.)

PLANTS THAT REPEL

Many pests locate their next meal from the scent a plant emits. For example, if you plant a plot of only broccoli and cabbage, you're making it easy for the cabbage moth! But if you surround and interplant that same area with carrots and onions, you mask the scent of the broccoli and cabbage. Those moths are going to have a hard time finding their meal, and they might fly off in search of someone else's garden.

Some plants contain phytotoxins that sicken or kill pests. For example, cabbage and similar plants contain mustard oils that kill flies, mosquitoes, spider mites, and Mexican bean beetles. This makes cabbages good companion plants for beans.

I've noticed another useful effect in my own garden: Flea beetles love to devour cabbage and cauliflower, but they never seem to bother the sticky, hairy leaves of tomatoes. When I plant tomatoes with either of the other two vegetables, the beetles bother none of the plants.

Additional repellent plants that may prove effective are leeks, onions, and rosemary against the carrot fly; parsley and tomatoes against the asparagus beetle; geraniums and petunias against leafhoppers; southernwood against cabbage moths; and nasturtiums against whiteflies. A Washington State University study indicated that sagebrush (*Artemisia tridentate*) actually turned on pest defenses in tomatoes.

Although some plants are repellent, they might take over your garden. For example, catnip can repel flea beetles, Colorado potato beetles, and green peach aphids but because catnip self-seeds, it has a tendency to get away from you. Instead, it can be grown outside the garden and cut and used as a protective mulch. Other invasive plants that have repellent effects, such as tansy, can be made into a tea and then sprayed on plants to repel squash bugs.

Sometimes a plant can repel larger pests simply by creating a physical barrier between the critter and the plant it wants to eat. If raccoons are raiding your corn, for instance, surround it with a scratchy barrier of squash vines.

PLANTS THAT NOURISH

Certain plant allies actually protect others from disease and improve the flavor of neighboring vegetables by providing nutrients. Think of herbs that enhance a vegetable's flavor in cooking, and grow those combinations together. Basil goes with tomatoes, dill helps cabbage, and summer savory is a natural with beans. On the all-vegetable front, carrots, green onions, radishes, and lettuce make great companions—and great salads.

Other plants can improve the health of their neighbors by protecting them from disease. For example, garlic and chives help prevent black spot on roses and scab on apples, and horsetail (infuse the plant and use as a spray) works against rust, mildew, and fungus. Summertime heat can take a toll on radishes, spinach, lettuce, and turnips. Here, larger plants, such as pole beans and tomatoes, provide shade, conserving moisture and reducing heat that would cause these vegetables to become woody or bolt, and produce flower stalks.

Some allies are grown as cover crops, used to prepare a garden. They bring up necessary nutrients that include trace minerals, or pull nitrogen out of the air and into their roots, which decompose back into the soil. For example, tomatoes need plenty of calcium, phosphorus, and trace minerals for flavor. Because buckwheat can supply all these, grow it as a cover crop the year before planting tomatoes. Along the same lines, nitrogen-fixing crops such as fava beans, alfalfa, vetches, and clover can be grown and turned into the soil before growing nitrogen-hungry plants such as squash, corn, celery, greens, and broccoli.

Companion planting is a practice long used by organic gardeners. Diversification is fundamental in any garden, and most experts agree that a mixed or interplanted garden is always healthier than an area growing just one kind of plant.

PLANT COMBINATIONS TO AVOID

Just as some plant combinations make good neighbors, others don't get along. Invasive plants must be separated from plants they'll overtake. Other plant pairings compete for the same nutrients or attract the same pests, lowering the defenses of both. Some plants release substances into the soil that can inhibit others.

Watch your own garden, and if some of your seeds won't germinate or some of your plants are failing and you've exhausted all other possibilities, consider the plant growing nearby.

Plant Combinations to Avoid

Plant	***Keep Away From***
Beans	Onions, garlic, shallots, leeks
Beets	Pole beans, mustards
Broccoli	Strawberries, grapes, sunflowers
Carrots	Dill
Corn	Tomatoes, sunflowers
Eggplant	Potatoes, tomatoes, peppers
Garlic	Beans, peas
Lettuce	Cabbage family, sunflowers
Onions	Beans, peas
Peas	Onions, garlic, shallots, leeks
Peppers	Eggplant, tomatoes, potatoes, fennel
Potatoes	Eggplant, peppers, tomatoes, squash, pumpkins, sunflowers
Strawberries	Cabbage family
Tomatoes	Corn, potatoes, peppers, eggplant
Turnips	Mustards, cabbage family
Rutabagas	Mustards, cabbage family

Beneficial Garden Companions

Here are some favorite plant combinations used by gardeners throughout the years. Test them out in your own garden and see how they work for you.

Beneficial Plant Combinations

Plant	*Highest Benefits To*	*Effect*	*Comments*
Basil	Asparagus, peppers, lettuce, tomatoes	Repels pests; improves health/flavor	Enhances growth, repels flying insects
Beans	Potatoes, beets, squash, lettuce, cucumbers, carrots, corn	Repels pests; adds nutrients	Deters potato beetles
Borage	Squash, tomatoes, strawberries	Repels pests; improves health/flavor; adds nutrients	Provides minerals and repels tomato worm
Calendula	Broccoli, cabbage, corn, beans, peas, spinach	Attracts beneficials	Attracts minute pirate bugs and lacewings
Catnip	Eggplant, oriental greens, potatoes, lettuce, peas	Repels pests	Grow separately and use as mulch
Chamomile	All garden plants	Repels pests; attracts beneficials; improves health/flavor	Improves crop yields, attracts parasitic wasps, ladybugs, lacewings
Chives	Tomatoes, carrots, apples, berries, grapes, roses	Repels pests; improves health/flavor	Deters Japanese beetles, aphids
Dill	Cabbage family, cucumbers	Attracts beneficials; improves health/flavor	Growth/flavor of cabbage family
Garlic	Tomatoes, cane fruits, fruit trees	Repels pests; improves health/flavor	Repels potato blight, also use as spray
Geraniums	Cabbage family, grapes, roses	Repels pests	Repels cabbage worms, Japanese beetles
Goldenrod	All garden plants	Repels pests; attracts beneficials	Attracts big-eyed bugs, ground and soldier beetles
Marigolds	All garden plants	Repels pests; attracts beneficials; improves health/flavor	Repels aphids, potato and squash bugs; long-term use kills nematodes

Plant	*Highest Benefits To*	*Effect*	*Comments*
Nastur-tiums	All garden plants	Repels pests; improves health/flavor	Deters many pests, attracts black flies
Onions	Most crops, except peas or beans	Repels pests	Deters many pests, especially maggots
Petunias	Eggplant, grapes, greens	Repels pests	Also plant with any vegetables bothered by leafhoppers
Potatoes	Bean, cabbage family, corn, melons	Repels pests; attracts beneficials	Repels bean beetles
Radishes	Beans, carrots, cucumbers, lettuce	Repels pests	Repels cucumber beetles
Sage	Cabbage family, carrots, tomatoes	Repels pests; attracts beneficials; improves health/flavor	Deters cabbage moths, carrot flies
Savory	Beans, peas, onions	Attracts beneficials; improves health/flavor	Generally beneficial
Tomatoes	Asparagus, carrots, cabbage	Repels pests; improves health/flavor	Tomatoes and asparagas are mutually beneficial
Yarrow	Most aromatic herbs	Attracts beneficials; improves health/flavor	Attracts parasitic wasps, ladybugs, tachinid flies

Test Your Soil

Cost savings

Up to $100 per growing season possible in improved yields

Benefits

Better crop yields

With the right ingredients, texture, and moisture, soil can bring forth improved crops of fruits, vegetables, and flowers with little monetary investment on your part. Here, Lisa Jansen Mathews shares her method of determining soil quality.

My farm is next door to the California State historic park called Malakoff Diggins. Malakoff is the site of the most successful hydraulic mining operation in the 1870s. Entire mountains were washed away with huge hydraulic monitors while the tailings (the flushed gravel and soil) were screened for gold. The soil was forever disturbed. Hydraulic mining was outlawed in the 1880s. However, the mining areas, known as diggins, still appear as moonscapes. My soil has been a challenge, but it improves each year as I continue to add compost, manure, and animal bedding.

When making piecrust, I can identify a missing ingredient by the texture of the dough. If it's sticky, I know I mismeasured the flour. If it's dry, I used the wrong amount of shortening or water. Soil texture gives me information, too. Soil that remains wet or cracks when dry tells me the clay content is high. Soil that will not hold nutrients may be high in sand. Soil containing hard clumps and dust may have too little organic matter. Ideal soil contains a mixture of sand, silt, clay with generous organic matter, balanced bacteria and fungi, and worms. Of course, I am being somewhat simplistic. Soil is chemically and biologically complex, but you can have productive soil without a Ph.D.

The health of your soil can make watering easy or a nightmare. Heavy, compacted, poor soil can allow water to run off, stand unabsorbed, or saturate to sogginess. A healthy soil composition with good texture absorbs and holds water while making it available to plant roots. Organic matter, such as compost, improves soil texture, allowing retention of moisture, and water content is critical to a healthy garden.

I knew my soil was going to be a long-term challenge. While analyzing and correcting years of soil abuse, I gardened in raised beds filled with soil purchased from a landscaping supply business. This interim solution allowed me to enjoy the rewards of gardening and take a natural approach to rebuilding my indigenous soil.

SOIL COMPOSITION AND TESTING

An easy way to find out what your soil is lacking is to test its composition. Into a quart Mason jar, put 2 cups (450 ml) of your topsoil; then add water, filling the jar. Tighten the lid, and shake the jar until all the soil particles are in solution. Allow the mixture to sit for 24 hours, during which time the sand, silt, and clay will separate. The sand will settle on the bottom, the silt in the center, and the clay at the top of the soil mass in the jar. Clay stays in suspension a long time in high clay soil. (Because my soil is very high in iron, I used a half-teaspoon of fabric softener to speed up the process of settling out the clay portion.) Above the soil will be water and maybe some organic matter floating on top. This will give you an idea of whether you have an even mixture or not.

An ideal loamy soil will be approximately 40 percent sand, 40 percent silt, and 20 percent clay. One solution to an imbalance is to add whichever element you're lacking. Check landscaping supply businesses for availability of sand, silt, and clay. Or, you can tailor your gardening practices to your soil type.

For example, my soil is considered a sandy-clay soil. It is roughly 50 percent sand and gravel, 45 percent clay, and 5 percent silt. I need to water more often than most because sand drains readily. However, the small particle size of clay holds nutrients well, and the sand content allows for good drainage, allowing me to grow plants requiring this asset. Each soil type has its advantages and disadvantages.

In addition to the sand, silt, clay, and organic matter in soil, there are soil nutrients and trace minerals. Determining the nutrient content of soil can be challenging. You can buy home soil-testing kits or meters, or you can send the soil to a testing laboratory. Many home soil-test kits test pH and nitrogen, phosphorus, and potassium levels for as little as $20 a kit. Soil-analyzing meters can take measurements of fertility, light, moisture, and pH. Meter prices range from $25 to $400. I prefer to work with a laboratory soil analysis, which usually tests for more soil ingredients. Many include levels of micronutrients such as manganese, iron, copper, and boron. Some labs or catalogs offering testing have books that aid in understanding your soil analysis. Prices vary. My soil analysis cost around $30, which is a small investment if you do any serious food gardening, because store-bought produce is so expensive.

SOIL TEXTURE AND AMENDMENTS

You can correct the soil texture by amending soil composition and adding organic matter. Compost is a rich source of organic matter and a great way to improve soil texture. You can buy compost or make your own. (See "Compost in Your Backyard" on page 222.) I use homemade composters, one made of a stack of old tires and another of an old water-softener tank. No matter how you make it, compost is an asset to your soil.

When soil composition and texture are good, soil moisture is easier to regulate. Good soil holds water like a wrung-out sponge, not soggy and not dry. Irrigation practices can enhance or damage soil. Overhead irrigation such as sprinklers or squirting with a hose can build a crust and compact soil. This method of watering can also erode topsoil, washing it away. On the other hand, drip irrigation conserves water, delivers water to the root zone, and protects soil texture.

Knowing your soil composition, adding compost, monitoring your soil's nutrients, and using good watering techniques make up the recipe for a successful garden. With a little work, you can have the perfect mud pie!

Index

H

I

J

T

About the Contributors

Gary F. Arnet is a retired dentist who lives in Northern California. He writes on outdoor living, preparedness, and self-reliance skills.

Mrs. John Bayles is a contributor to *Back-Home* who writes on a variety of practical farm-stead subjects including gardening and cooking.

Sandra K. Bowens shares her cooking and camping recipes from the mountains of Colorado.

Tim Burdick is a good steward of the earth and serves a ministry in Tacoma, Washington.

C. E. Chaffin is a home gardener who writes on the benefits of composting from the Heart of Dixie.

Gretchen F. Coyle and her husband, John, are organic farmers who share the value of found mulch. Gretchen writes from coastal New Jersey.

Susan Dabuiskis-Hunter contributes to *BackHome* on topics of ecological responsibility. She lives in southwestern Ohio.

Lori Dzierzek writes for *BackHome* and shares her practical experience and wisdom from a small farm in rural Virginia.

Thomas J. Elpel is a sustainable designer/builder and outdoor enthusiast, and the founder/director of the Hollowtop Outdoor Primitive School in Pony, Montana. He is the author of five books on building and natural living (www.hopspress.com).

Kim Erikson is a contributor, gardener, and home canner writing about food preservation from Southern Nevada.

Becky Flatau is a retired elementary school teacher who enjoys cooking, gardening, and sewing. She writes on crafts from Merrill, Wisconsin.

Richard Flatau is a special education teacher, gardener, and cordwood masonry advocate. He is codirector of Cordwood Construction with wife Becky in Merrill, Wisconsin.

Catt Foy is a sustainable living enthusiast writing from Davenport, Iowa.

Heidi Gaschler is a contributor from western Kansas who enjoys working outdoors with animals, art, cooking, reading, playing the piano, and living in the country.

Dottie Goudy is an avid gardener and shares her no-cost mulching techniques from the Gulf Coast of Texas.

Susan Grelock writes on whole foods and grains from Northern California.

Marie D. Hageman is a New Jersey gardener who combines stonescaping with water and soil retention to save money and make her plantings more productive.

Nancy Hamilton is a contributor and natural fiber artist who currently lives in Tennessee.

Ann Heatherley shares her memories of a beloved family member and a simple homemade doll.

June Higgins is a *BackHome* contributor, pet lover, and devoted seamstress for her dog Beedj.

Maggie Julseth Howe is co-owner of Prairieland Herbs in Woodward, Iowa. She writes on crafts and natural health for *BackHome*.

Cynthia Hummel is an outdoor enthusiast who shares her recipes for safe and inexpensive herbal insect repellents.

Bill Janes is an avid gardener and the former advertising director for *BackHome* magazine.

Judy Janes is the food editor for *BackHome* magazine, where she develops and tests recipes for photography and publication.

Kellie Janes is a teacher and outdoor enthusiast who lives with her son and husband Matthew in the Bahama Islands.

Shawn Dell Joyce is a sustainable artist and activist in New York State's Hudson Valley, where she founded the nonprofit artist's cooperative the Wallkill River School and writes a nationally syndicated environmental awareness newspaper column.

Pat Kerr is a contributor to *BackHome* and writes about gardening and country living matters. She lives with her husband in Ontario, Canada.

Gerald Y. Kinro writes on a variety of topics from the island of Oahu, Hawaii.

Joyce and Jim Lavene are a married writing couple from North Carolina. Together they have more than 50 books in print.

Karen Lawrence is a winter gardener who shares her low-cost season-extending tips from a variety of sources.

Kessina Lee-Wuolet harvests rainwater and contributes from northern Oregon.

Beth Lorms writes on kitchen and cooking from central Ohio.

Greg Lynch is a natural foods advocate in Minneapolis, Minnesota

Lisa Jansen Mathews writes for *BackHome* and contributes on practical homestead topics. She lives on a small farm in rural California.

Lynda McClanahan is a *BackHome* contributor who saves money making homemade soaps.

David McCormick crafts chairs from found materials in the vicinity of Springfield, Massachusetts.

Beverly K. Metott is a gardener, heirloom seed-saver, and frequent contributor to *BackHome*, writing on topics of rural living.

Mary Ann Osby is a public school teacher and yoga practitioner in western North Carolina.

Darlene Polachic is a book author and professional writer and contributes on a wide variety of subjects from Saskatchewan, Canada.

Merwyn Price is a workshop handyman who shares his must-have tool choices with those not in the know.

Nancy Reiss is a veteran New England gardener and explores the advantages of raised-bed gardening.

Anna Lauer Roy is an outdoor guide, natural builder, and whole foods advocate. She lives in Colorado.

Patricia Rutherford contributes on topics from western Illinois.

Jane Scherer is an herb and garden enthusiast who shares her recipes with *BackHome* readers from her home in Long Island, New York.

Louise Schotz is a retired high school science teacher, now a professional craftsperson working in textiles, quilting, beading and jewelry. She is the recipient of several awards and teaches arts from her Outback Studio in Irma, Wisconsin.

Shawn M. Schulz is a *BackHome* contributor and writes on practical home animal care and country living.

Marcella Shaffer is a contributor who covers everything from food preservation to livestock to herbal vinegars. She writes from northern Montana.

Charlotte Anne Smith is a *BackHome* contributor who shares her experience in home organization and uncluttering from her home in Oklahoma.

Sandra Noelle Smith writes on herbal recipes for *BackHome* magazine.

Lynn Smythe is *BackHome* contributor whose area of interest is herbs and spices. She lives in south Florida.

Garth Sundem is an outdoor enthusiast and contributor to *BackHome* magazine. He lives in Bozeman, Montana.

Sara Trachtenberg writes on natural health and beauty from the Boston area.

Charles Dickson, Ph.D., is an ordained minister and college chemistry professor and author of more than 500 articles and nine books. He is a *BackHome* contributor who specializes in natural health and healing practices.

Kris Wetherbee (www.kriswetherbee.com) is an internationally published author, writer, and speaker specializing in food, gardening, nature, and outdoor living. Her articles have appeared in more than 85 magazines. Kris and her photographer-husband, Rick, enjoy developing and maintaining their gardens in Western Oregon.

John Wilder is a freelancer who writes tool reviews and construction projects and specializes in green and environmentally sound home improvement techniques. He lives in Jacksonville, Florida.

About the Author

Richard Freudenberger is the technical editor of *BackHome* magazine and the author of a number of books on woodworking, home maintenance, and renewable energy. He lives in the rural mountains of western North Carolina, where he keeps bees and chickens and develops sustainable living skills.

BackHome Magazine
P.O. Box 70
Hendersonville, NC 28793
800.992.2546
www.backhomemagazine.com